"Uniquely useful for graduate education, Ruopp offers a framework for exploring how visual tools can help make sense out of dense theoretical readings, often required for masters and doctoral students about to engage with research. Encounters with Theory is an invitation to rethink the power of the visual and how new understandings of complex theories can emerge through different modes of visualization. Deeply grounded in her practice as an artist, teacher, and researcher, Ruopp writes with candor and humility, clarity, and intelligence. Follow her thinking into collaboration with graduate students. Learn about generative, creative, multidisciplinary, and even uncomfortable processes can lead to new insights. Muse over visual evidence as well as verbal exchanges entwined with theoretical readings. As Ruopp asserts, researchers have much to gain when the verbal and visual are afforded equal billing in the pursuit of data analysis, knowledge acquisition, and new ideas."

Karen Lee Carroll, *Ed.D, Dean Emeritus, Center for Art Education, Maryland Institute College of Art, USA*

Encounters With Theory as Conceptual Medium and Creative Practice

Encounters With Theory as Conceptual Medium and Creative Practice explores the relationships and intersections between verbal and visual ways of researching, challenging the privilege of the written word in academe.

Rooted in a grant-funded artistic research course, the data and experiences shared here illuminate the transformative power of visual thinking and visual literacy as a research data, analysis as well as artifact. The book begins by outlining the author's background as an artist/researcher/teacher, laying a foundation for the positionality and thinking within the book. The later chapters, offered as vignettes, share the explorations and subsequent discoveries of emerging scholars from a variety of backgrounds and disciplines. A/R/Tography takes a front seat serving as a messy and fluid architecture for theory put into practice. Engaging tension as a catalyst for disruption, the book explores how staying present, intra-acting with ideas, concepts, and theory through visual material exploration expands attention and illuminates data in different ways, affording unexpected insight and discovery. In addition, the image-rich pages invite readers into reading the visual in conversation with the verbal, on equal terms. One is not in service of the other, rather a conversation of literacies.

With its emphasis on the visual and materiality as a method of thinking, this book will be of interest to anyone interested in or practicing artistic research. One does not need to have an arts background to engage in visual dialog.

Amy Ruopp is Associate Professor, Chair of Art Education, and Assistant Dean of Undergraduate Studies at the College for Creative Studies, USA.

Encounters With Theory as Conceptual Medium and Creative Practice

Amy Ruopp

Routledge
Taylor & Francis Group

LONDON AND NEW YORK

First published 2023
by Routledge
4 Park Square, Milton Park, Abingdon, Oxon OX14 4RN

and by Routledge
605 Third Avenue, New York, NY 10158

Routledge is an imprint of the Taylor & Francis Group, an informa business

British Library Cataloguing-in-Publication Data
A catalogue record for this book is available from the British Library

ISBN: 978-0-367-42340-7 (hbk)
ISBN: 978-0-367-42343-8 (pbk)
ISBN: 978-0-367-82381-8 (ebk)

DOI: 10.4324/9780367823818

Typeset in Times New Roman
by Apex CoVantage, LLC

To the childlike curiosity that lives in all of us!
May this reignite and renew an authentic sense of
wonder to wildly explore possibilities free from the
constraints of what you have already learned.

Contents

Acknowledgments

What is any composition of words and images without the countless moments, people, and experiences that collectively make them? This book is a long time coming and could not have been completed without the patience of Hannah Shakespeare and all the people at Routledge! Two years of a pandemic really makes time and progress slippery.

The chapters of this book contain people and ideas that animate thought and represent a fearless devotion to play, curiosity, and a surrender to the messy. Special thanks to the students who went on this journey with me and especially those who allowed me to share their own live inquiry in the chapters of this book. Likewise, my partner in all things creative, Dr Kathy Unrath who supported my journey through the co-writing of a research grant which afforded me the opportunity to write and teach the course which provided the data for this project. She then participated in the course as both a participant and a co-facilitator. It is everything to have someone who unconditionally believes in you while challenging the norms of academe. There were many others who supported me in this process, too many to name, but a few require a public expression of gratitude; Dr. Karen Carroll and Dr. Sharon Johnson, my long-time mentors and muses who read and offered insights on my early drafts. My amazing sister, Stephanie Ruopp, who is a writer in her own right and a wizard when it comes to editing! My daughters, Sage and Veyda, who tolerated my frustrations along the way and my little dogs Roo and Ellie who slept on my lap as I struggled to find words. I think in images before words. Images inspired the words and narratives on these pages. Finally, my gratitude extends to the physicality of material as we all engaged in a co-composed dialog of inquiry, messy discovery, and the visual mattering of theory.

A Foreword Looking Back

The essence of this forward is forward looking back (that (in)between space between then and now). It is a privilege to look forward as I recall where this journey began. I met Amy Ruopp 30 years ago when she entered my middle-school art classroom as my student teaching intern. We both shared the view that art education is more than learning to be a creative producer but to emphasize the significance of the process of creating, respecting our middle school students' ability to investigate and interrogate their own creative process and growth. We agreed that art education is also about facilitating deep noticing of the world around them, the ability to see and perceive, and to read and decode the world they live in. Our mission was to bring our students to what C.K. Chesterton (2017) called the sunrise of wonder. We shared a commitment to our profession and a desire to expand the horizons of our field and to use the precious time we had with our students to generate curiosity and to encourage our students' creative potential and meaning making. Our time together was magical!

Our paths intersected again when Amy began her doctoral program with me at the University of Missouri. The imaginative sparks again began to fly and along the way we taught, investigated, published, presented, and created visual art together as a dynamic duo of intrepid art educators. We recognized the dynamic of teamwork and conducted visual research into the essence of mentorship to better understand our own alchemy. One of the most profound insights to come from our mentorship research is that "we make each other better" (Unrath & Ruopp, 2017). This symbiotic intra-active connection is significant to understanding how we moved forward to the artistic research course described in this book. As I look back, I can see the seeds of this artistic research course through the lens of the myriad conversations and discoveries that we shared over 30 years which intensified and strengthened during Amy's doctoral studies. We are both a/r/t/ographers and thus share an understanding that each of our identities (artist, researcher, and teacher) are entwined, fluid, and inextricably connected within ourselves but together our voices melded then and now.

During my career as an art education professor at an AAU research intensive university, I strived to model artist/teacher practices, develop those new and incoming to our field, and sharpen the intellect and perceptive skills of those who wish to contribute new voices to the field through research in art education. Soon I recognized that artistic research is not readily understood, visible, or accessible to students across the disciplinary spectrum of university research coursework. I wondered if those in the academy ever considered what artistic research entails, whether it has a methodological basis or construct that aligns with their conceptions of what academic research can or should be. I recognized that there was a void in our educational landscape and thus, saw an opportunity to design and develop a vibrant and engaging course that could shed light on how artists create and the intra-active influence of visual/verbal literacies.

In 2016, Amy and I began to discuss the possibility of a research course that could transcend traditional disciplinary boundaries and cross differing research paradigms to engage with visual literacies in an integrated research/studio space invigorated by a sense of creative play and self-discovery. Based on our early mutual understandings about art education, Amy's doctoral research, and our prior collaborative research, we believed that engaging students in a multimodal reflection on the art-making process could produce a profound and nuanced personal understanding of the process of art making (Ruopp & Unrath, 2019). Fundamental to the development of this course was Amy's recognition of the significance of visual literacies and artistic research as a way of knowing and investigating a world of symbol systems as data. Amy's design for this course put to work her previous experience and research as she strategically positioned the course in a studio space to put emphasis on the visual as valuable data for research while also disrupting the privilege of the written word. Our ideas about making learning visible were thus applied to making research visible and understanding what artistic research could produce.

As this course commenced in the Fall of 2017, I eagerly attended every class, read the assigned readings, and enthusiastically participated in every class activity. I was soon transported into new encounters with theories, text, images, artists, and students from different disciplines and educational backgrounds. I interacted with puppet makers, critical family historians, artists, and art educators. Each week we came armed with visual representations of our learning in image and text. There was an anticipatory excitement about what we had newly discovered and created and an expectation that we would DO something new and wonderful with this information in class. These artifacts held what Bennett (2010) calls "thing power", "the curious ability of inanimate things to animate, to act, to produce effects dramatic and subtle" (p. 6). Thus, our visual/verbal products contributed to our discussion as they were considered by our group.

Our new way of deep reading/drawing/seeing/feeling/re/reading our assignments created a new comprehension of the material as our perception was slowed down and revisiting text and image was encouraged as a new practice. Each class, we considered the work of select artists (I was very honored to be included as a featured artist) and weekly we investigated new media in a space that encouraged playful exploration. Each student's transformation brought new epiphanies as students moved from discomfort to insight, from playfulness to discovery, and from apprehension to courageous personal breakthroughs. There was an exciting atmosphere of creative inquiry where thought fireworks created sparks of perception and self-awareness. It was electric.

Now, three years out, Lauren, then an art education doctoral student, reflects on her own trepidation at the beginning of the course:

> *I can still remember those first few classes, sitting in the college art studio (the same one I had sat in almost two decades before) exposed and powerless, unable to put meaningful marks on my paper, questioning my ability to 'be' in academia, and uncertain about this journey I had embarked upon. For months, I wrestled, fought, and collided with an entanglement of ideas and identities as I trudged through post-structural philosophy, the landscape of a/r/tography, my own lack of confidence and sense of self, and the demands of what the academy required. Being exposed and immersed in this untraditional way of thinking, knowing, and doing was in stark contrast to the long-established norms of qualitative research I experienced simultaneously. It left me depleted, disrupted, and with a sense of such astounding wonder.*

Epistemologically, I fundamentally believe that all new learning comes from discomfort, that it takes effort to push beyond what you believe to create anew, to know and grow, and to imagine, disrupt/reimagine, and re/create. As the semester progressed, we all relaxed into the discomfort of wrestling with new ideas that confronted our belief systems and settled into experiences with the understanding that as a community of learners we were all making each other better.

As you read this book, you will meet Amy more fully as an artist, researcher, and teacher. In Chapter 2, Amy shares the theoretical and conceptual framework that undergirds this course design and implementation and in Chapter 3, she provides a detailed explanation of the course rationale and design with a weekly outline of class activities and assignments.

In Chapters 4–7, you will meet four amazing students whose insights and transformational musings are described below:

Gina, a doctoral student in Theatre, continued beyond the course to pursue an independent study with Amy as she prepared for her dissertation proposal. Now, looking back, she shares that she found a home in this new community:

> *As a visual thinker, I always felt somewhat of an outsider within the rules of academia, especially when engaging in traditional forms of written knowledge and research. My experience in the Artistic Research course helped me realize how my unique interpretation of learning, doing, and understanding of the world could be an asset and not a deficit to my work. This course offered me the tools I needed to reimagine research that drew from my various skills as a visual artist. I also found a new community of visual thinkers and mentors to support me throughout my academic career.*

Brooke, an art education doctoral student who holds an MFA degree, was profoundly impacted by the nuanced theorizing that she experienced:

> *The thinking and creating that occurred throughout this course greatly impacted my conception of what it means to conduct research. I was given permission to play, to not be so sure, and to embrace the unknown. Research is messy. It does not exist in isolation. It is always unfolding, intra-active, and emergent. To intentionally seek out entanglements and disruptions was a novel experience for me—not to put things back together, but to move with and in-between the rough edges and the instability. Research and artistic practice became equal agents in motion with and because of one another. Through the intra-actions that occurred in this course, I interrogated my previously held practices about creation and binary avenues of thought. As a result, my artistic research became simultaneously intentional and unexpected, tense and fluid, and ultimately an ongoing, open-ended practice.*

TWK was drawn to this course as a way of performing her research in Social Studies Education and a desire to develop and understand her creative spirit as well as the theory that supports her thinking. The course was enhanced by the cross-disciplinary strengths of our community as she, in retrospect, discusses her artistic processes and validation of her own artistic responses:

> *I uncovered two parts of me lurking in the margins*
>
> *When I began this artist journey in the middle stage of life, I uncovered two parts of me lurking in the margins: the long-ago little girl and the woman-to-be; both becoming simultaneously.*

Figure F.1 Shade and light

Hidden away was an energetic curiosity that needed space to explore. This vibrancy subsequently produced a series of "styk people"—scribbled beings with thing-power who desired to roam freely and escape the stifling binaries and gruesome shades of habit and tradition.

These were the pent up and unimpeded spirits that escaped my daydreams and imagination. They demanded to be emancipated from the predictable, worldly, and oppressive space and comfort of sameness which masqueraded as normative lullabies.

On some days the knowledge of both selves worked together, attracted by sound, colors, images and melancholy. The styk people carried the harmony and the disassociation with mutual regard. Other times, the past and memories kept them apart, a psychological necessity for me, being a woman and being black, operating as a mother, friend, and wife. The researcher part of me, the academic

Artistic inquiry gave me both shade and light. Purpose and possibilities converted letters and syllables into images that haunted and poured the affective spirit of my imagination's surplus, the excess weight of what I could not see, into a subversive form of play and knowing.

She's still here and there and that is ok with me.

We had a vision of a space where ideas could be seeded, concepts could be refracted and entangled, and theoretical and personal growth could be nurtured. Ultimately, every teacher aspires to the hope that the garden they cultivated would resonate in a deep and thoughtful way with their students and that the blossoming of thinking sown in that liminal space will continue to root and spread and thrive. Indeed, echoes of these experiences have impacted the trajectory of successful doctoral research as Lauren illustrates here:

> *As I sit here, a newly minted Doctor of Philosophy, I can say with certainty this methodological madness changed the trajectory of my existence. This existence is now rooted in an onto-ethico-epistemological understanding of the world, a knotted understanding of myself as an artist, researcher, and teacher, a troubling of how one 'does' research, and as MacLure (2013) so beautifully conceptualized, a wonder that arrives unannounced because of our ability to notice and exist in between the spaces of the known and unknown. This is the space where I am most at home.*

Dr Kathy Unrath
Associate Professor of Art Education Emerita
University of Missouri

References

Bennett, J. (2010). *Vibrant matter: A political ecology of things*. Duke University Press.

Chesterton, G. K. (2017). *Autobiography by GK Chesterton-Delphi classics illustrated* (Vol. 79). Delphi Classics.

MacLure, M. (2013). The wonder of data. *Cultural Studies-Critical Methodologies, 13*(4), 228–232.

Ruopp, A., & Unrath, K. (2019). Building theory: Making artistic learning visible. *Visual Arts Research, 48*(2), 29–48.

Unrath, K., & Ruopp, A. (2017). Cultivating potential-harvesting wisdom-an a/r/tographical illumination of mentorship. *Journal of Visual Inquiry, 5*(3), 433–447.

1 A Composition of Moments

How does one begin something that has already started? The content of this book has been assembling for some time. The notes, charts, drawings, experiments, long walks, and daydreams, weaving select words mined from my encounters with theory, research, curriculum, art, and the inter/intra-actions (in) between all of them are mattering, quite literally, here. This could have been composed a million different ways (well, within the limitations of the prospective publisher, of course). Knowing and feeling that was paralyzing for me, until it was not anymore and I jumped into a section. Not the one you are presently reading if you started, well . . . here.

Some of the contents of this book emerged from countless encounters co-composing (Manning & Massumi, 2014) with others through reading, teaching, and creating. Yet, much of the data originate in my life as an artist, my passion as an educator, and my insatiable wonder as a researcher. In re/searching my own historical narrative, and sharing a slice of it here, my intention is to offer in/sight as to my own lens and context. However, it need not be the one you adopt. I hope you bring your own lens to this and add to the offering here.

Roots

For me, drawing and visualizing have always been a way to discover what I want to know, and more importantly, a mechanism for revealing what I did not know. Even as a younger adult, I believed I did not know how to write. During my elementary years, I was put in the "dumb group" for spelling and language arts. It was humiliating. Quite honestly, that experience contributed to the way I read the world so visually today. Being placed in remedial groups early on my educational path was one of the first signs that my way of navigating knowledge was not valued and, taken further, marginalized. As hard as that was, I was resilient and persevered, developing my visual

DOI: 10.4324/9780367823818-1

Figure 1.1 Images whisper words

skills and spending countless hours drawing, illustrating stories, and living in make-believe worlds that engaged all the senses. I found myself doodling through class lectures. It helped me to focus and make sense of what was being shared. Sometimes, I got in trouble for what might have appeared to be not paying attention. The truth was, drawing focused my attention and facilitated comprehension. After high school, I went to college to become an artist, followed by an art teacher, and finally, a teacher of art teachers. All of this visual work landed me in the space I am in right now; which is ironically struggling to **write a book**. I came across this quote by James Baldwin and it resonated with me;

> *When you're writing, you're trying to find out something which you don't know. The whole language of writing for me is finding out what you don't want to know, what you don't want to find out. But something forces you to anyway.*
>
> *(James Baldwin, unknown)*

There is tension in this quote and if there is one thing I have learned along this journey, it is that tension is opportunity. In addition, the sentiment of this quote very much echoes my experiences with my own visual creating and knowing. My conceptualizations of creating are that **the process is the product**. Ingold (2010) points out that artist "Paul Klee repeatedly insisted that the process of genesis and growth that gives rise to forms in the world we inhabit, are more important than the forms themselves" (p. 91). Process informs, and deliberate attention to it keeps one in the moment of unfolding realization. The ideas and experiences of my past, present, and future are important. How I make sense of the world literally matters through my visual thinking. I must claim my artist-self as a scholar, one who encounters knowledge through multiple literacies.

I KNOW I am not alone on this path. I have encountered brilliant thinkers who express knowing in and through visual ways. Artists of course come to mind, but so too do many others who do not identify as artists. Mapping, charting, collaging, doodling, and graphing are all forms and expressions of thinking and knowing. These collective experiences celebrate the visual way of thinking and learning and inform knowing and the sharing of understanding in robust and rigorous ways.

In conceptualizing this book, I wanted the verbal and visual to be afforded equal billing. It is my intention that the linear quality of the words here serves as a bridge to the nonlinear aspects of visual thinking that is offered here as data analysis, knowledge acquisition, and emergence of new ideas. Image and text are in service of one another as equals. They should be read through and with each other, with curiosity and wonder! Allow yourself to surrender to wonder and wander leisurely through the different forms of texts offered. You might be surprised at what you discover.

It is important that you, the reader, understand I am a self-proclaimed lifelong A/R/Tographer; an Artist/Researcher/Teacher. This trio of identities is not a noun but a tangle of verbs. These identities are in action; they are doing, making, seeking, and exploring. They are in service to one another, through one another, etc. Irwin (2004) offers that a/r/tography as a practice or methodology invites all of us to "live a life of deep meaning enhanced through perceptual practices that reveal what was once hidden, create what has never been known, and imagine what we hope to achieve" (p. 35–36). I will speak more in-depth about A/R/Tography in Chapter 2 and use the term throughout the book as it is a conceptual framework for inquiry. It is interwoven in the content of every chapter, offering endless entry points into engagement. As an artist/researcher/teacher, the dominating lens of my trio is historically rooted in the artist. My earliest memories include my parents telling me that "someday" I would be an artist, as if children are not already. (Coincidentally, I wanted to be a lawyer as a child; given my penchant for questioning everything.) There will be numerous times I refer to artistic and creative processes in this book. Let it be known that I am not speaking for every artist. I can only offer insights into my own process and what that process has produced, specifically in the context of teaching and research here.

It should also be noted that there are many instances where other theorists, philosophers, researchers, objects, and teachers are invited into the conversation as they echo and validate my own experiences. "One never writes alone . . . one writing alone is already a crowd. Our [my] words in this book are never without the echoes of those whose difference we[I] chose to write with" (Manning & Massumi, 2014, p. viii). I am sure there are many voices missing. We are part of a collective experience. Because someone else wrote something before I do here, it does not make what I write here more or less valid. My words are not appropriating someone else's ideas or experiences because what I share here was lived authentically in my own original and unique way. With that, I apologize in advance to anyone reading this assemblage of literacies who feels they had the idea first or wrote about it and are not receiving appropriate citations. It is unintentional. Elizabeth St. Pierre and Karen Barad are the voices I converse with often. I have followed St. Pierre's advice of "read, and then read more", but at some point, there is only so much I can read until I have to 'do'. I am also a painfully slow reader, so it is in the doing that the reading is realized. For me, the doing is visualizing, intra-acting, and assembling verbal and visual texts in the same space to disrupt. I have been doing this for a long time without realizing it. This became apparent to me as I reviewed my own history as an artist/researcher/teacher.

A Brief Visual and Conceptual Retrospective

Looking and reflecting back over 30 years, my visual work has always been in service to something beyond the work itself. My early work was rooted in notions of manifestation. My focus was on how things became

Figure 1.2 Early work and seeds of inquiry

Figure 1.3 Manifesting vision

material, or mattered, with/in/to the world. Explorations were intentional, a dialogue with possibility. I was exploring topics and subjects that were close to me and the people with whom I was connected. At the core of this earlier work was making sense of the complex manner in which intersecting events unfold. Looking back, I realize the seeds for my current research and inquiry were already at work. The small faces present in much of my early work embodied the idea of possibility, intention, emergence, time, becoming, and unfolding. The essence and potential of these actions were conceptually incubating, waiting . . . The themes present in my current academic and creative research were seeded in the past. While thinking theoretically

Figure 1.4 Emerging ontology

Figure 1.5 Learning from glass

was not a conscious emphasis back then, theory had been inviting me, teasing me, and revealing itself through my continuous creative practice as a fluid constant with/in a rhizomatic tangle over time.

In 2001, I was introduced to hot glass and torchwork by a friend and was immediately consumed by it. My explorations in glass unfolded concurrently with my large manifestation drawings. There is an elegance to the notion "ignorance is bliss." I was self-taught and, I would dare say, the glass itself taught me. Here again, a seed, notions of posthuman theories, and object agency were working on my attention. Encounters which were shaping my future research were revealing themselves to me. Not knowing the rules and intra-acting with a material that was alive and expressing its own agency throughout the process gave way to experimentation unencumbered by rules and perceptions of how something was supposed to happen. While my drawings were exploring the mattering of intention, flameworking was about my relationship with material, thing-power. "Thing-power, according to Bennett, is 'the curious ability of inanimate things to animate, to act, to produce effects dramatic and subtle" (Jackson & Mazzei, 2016, p. 98). My relationship with glass was a sensorial dialogue between fire, glass, and the human exploring its nature. Or was it exploring mine?

Working with glass inspired a curiosity around the unfolding of ideation and the creative process. How are discoveries made? What does it mean to have an idea? I found myself preoccupied with moments of tension and decision as I created. My drawings were large, sometimes up to 10-feet wide and 6-feet high. I did not have the luxury of a large studio space that enabled me to step back and encounter what I was working on, so I took photographs along the way. This served several purposes. As I took note of how imagery emerged, I became much more aware of my own process and how I explored the surface and worked through tensions. **I realized tension and discomfort were my friends and inspired, if not demanded, growth.** I also appreciated the lens of the iPad and camera. There was something transformative about seeing my work through a different lens, scaled down. It offered a different way of responding to the emerging image and disrupted

Figure 1.6 Capturing process 1

Figure 1.7 Capturing process 2

Figure 1.8 Capturing process 3

Figure 1.9 Capturing process 4

fixed ideas I might have had about what I was drawing. Seeing the imagery from multiple different perspectives and sizes often compelled me to make different choices. I was flirting with the notions of diffraction without realizing it.

The realization that tension played a key role in my creative process began to uncover the theoretical construct of the rhizome and invisible connections. An important motivation that is ever present in everything I do is that of pushing boundaries, questioning, exploring, and wondering. Why? What if? I have become uncomfortable in comfort. When something becomes too familiar, I feel the need to interrupt and push beyond it. Koro-Ljungberg (2015) echoes my stance on tension when she discusses the idea that "interruptions also serve as examples of uncertainty and unthinkable energy" (p. xvii).

There is power, growth, evolution, and especially, potential in the spaces of tension. Creating in that space pushes me beyond what I think I know, disrupting habit. It is not unusual for me to suddenly pick up a brand-new material and explore its possibilities, allowing it to guide me as if I am the material seeking its own expression. Taking time to listen and sitting in

Figure 1.10 Solo show "Convergence"

what Patel (2016) describes as unsettled pauses are productive tensions and thresholds for growth.

In 2015, as part of my doctoral journey, I had a solo art exhibition. The title of the show was called "Convergence". While it did not have my glass in it, it did feature over 20 works created over a span of four years. Having all the work up in one space was a profound experience for me. It was as if the works were conversing with each other, creating invisible whispering interconnected tendrils through space, intersecting like a rhizome with echoes of the past converging in the present. "The rhizome as an architecture for thought offers a non-linear structure to explore both identity and process in action. The structure allows for multiple possibilities to emerge within an encounter, offering new perspectives and conceptualizations about creative practice" (Ruopp & Unrath, 2019, p. 34). For me, the space of the gallery conceptually became an invisible rhizome of theoretical ideation across time and space.

During this time, I was navigating the complex world of academe at a high research-intensive university. I was quickly confronted with the reality which traditionally privileges the written word. My brain was exploding. Flashbacks to my elementary school days haunted me. I struggled to read scholarly academic texts which were packed with deep meaning and rich with theory, ALL IN WORDS. In my frustration, I discovered that I have to read differently. For me, the only way to make sense of dense theoretical reading is to pause in my unsettled state, sink into it, and draw things out. Thankfully, I had understanding professors who honored my ways of knowing. Through visualization, I could articulate and make sense of what I was reading. It started with simple drawings along with my notes.

This quickly transformed into collages, combining imagery with text. By **slowing down** the reading process, I surrendered my fear of "doing it wrong" and cultivated an envisioning process for understanding complex theories. This led to visualizations of data and charts which enabled me to thoroughly explain and articulate my thinking in accessible ways to

Figure 1.11 Thinking through theory with imagery

Figure 1.12 The blending of literacies produces a more profound understanding

other nonvisual people. It also facilitated a more profound engagement with theory as I dove deep into my own research questions. I pushed myself to think about how scholarly research unfolds and I began to question the more traditional notions of research methodologies. Why are visual modalities marginalized, excluded or, at best, only supplemental to the written word? Another seed was planted.

I soon found myself cutting up articles and books to read in a nonlinear pattern. I was reading diffractively, reading one literacy through another in a

Figure 1.13 Mapping the movement of theory

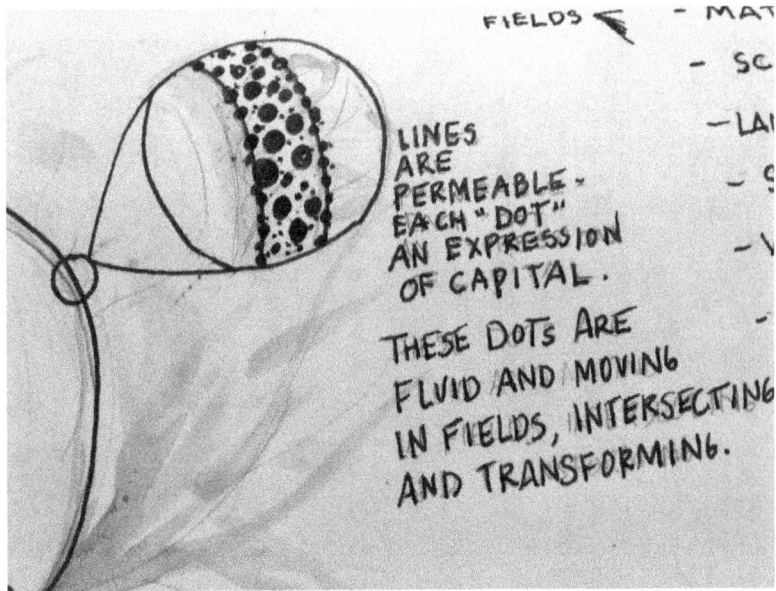

Figure 1.14 Details of mapping

nonlinear manner. Reading backwards, stopping, eliminating, crossing out, drawing on top of, and rearranging. I was engaging with traditional writing by reading it much the way I engage with my artist materials in my studio.

Because I believe all things are connected, this pushed my artwork and thinking in new directions. I saw and lived EVERYTHING as a creative act. I began to video myself working in my studio, capturing myself talking during imagery development. As I read theory, I made sense of it through artistic practice as I creatively explored it on a larger scale. Theory was literally mattering in my studio. Verbal and visual literacies were dialoguing with each other in the space of my studio as I, a conduit of sorts, moved in between them. My mark making evolved, color appeared, and written text started appearing in my work. Visual and written texts were comingling on equal ground. Everything was interconnected, and there were intersections everywhere! It was as if a giant invisible rhizome was entangling everything within and around me. All of this became data and simultaneous analysis for my dissertation work, and a diffracting mechanism for how I conceptualized and articulated art as a research act.

My current work continues to evolve and operate through the three lenses of the artist, researcher, and teacher as I explore the nuances of posthumanist theories and their intra-actions within the material world. Art and writing are not static. Both are an accumulation of intersecting moments throughout time. Both do something. They challenge the way we think, and

Figure 1.15 Literacies comingling in my studio

Figure 1.16 New mixed media work and pushing boundaries

each grapple with difficult questions while offering insights into the world we inhabit. They invite difficult dialogues while simultaneously educating. I believe Art is every bit as powerful a form of literacy as writing though and it is this that thrives at the heart of my research and teaching.

Pedagogically Emergent: Further Context of Coming Attractions

My own experiences as a visual and rhizomatic learner forced me to adapt and innovate beyond an academic norm that did not readily support the way I experience and inter- and intra-act with knowing. As a scholar, this created and continues to generate tensions in my research and content delivery. The

Figure 1.17 Text as image blends with visual to deepen meaning

visual aspect of knowing and understanding is often dismissed or misunderstood in a verbal academic world. Meanwhile, in the art world, artists have all been thinking and practicing this way for years, remaining mysteriously at the edges of or (in)between spaces of academic discourse. In my 30 years as an artist/researcher/teacher, it has become abundantly clear that I have response/ability to contribute to the blurring of boundaries between verbal and visual literacies. "My unique way of thinking has revealed a deepening approach to research. In thinking with and through a combination of literacies and theories, I might as Deleuze and Guattari (1987) suggest, realize new realities. . . . I cannot explain to anyone how things are rather, I explore how things become"(Ruopp, 2019, p. 2).

What follows in the coming chapters are musings of theory and practice and how new understandings *mattered* through different modes of visualization. Illustrated through vignettes of some brave students who participated in the Artistic Research course I designed and instructed (outlined in detail in Chapter 3), each student's unique experience offers possibilities around disrupting habits and traditional ways of knowing. These students came from multiple backgrounds and disciplines including Social Studies, Art Education, Fine Arts, and Theater. They surrendered wholeheartedly to the encounters offered in the curriculum. "When we blend such different ways of thinking, of being, of educating, of revealing truths, we engage hearts and minds simultaneously, in doing so making change in the world" (Walrath, 2017, p. 128). For these students, it likewise catalyzed change in self-perception, and their own relationship with questions, modes of inquiry, and research.

Chapter 2 begins by exploring A/R/Tography's role as a framework for thinking with and through theory and offers glimpses into the author's playful conceptualizations. A/r/tography opens spaces for sensuous discovery through material exploration. This form of inquiry centers on the process of finding rather than an emphasis on solving. A/r/tography, while rigorous, offers unique platforms for investigation honoring the unique lived experiences of each artist/teacher/researcher. As a creative framework, it offers multiple entry points into experience. Receptivity to ambiguity, the disrupting of language as it becomes entangled with image, material intra-actions and the tensions these moments produce offer a reconsideration as to how research unfolds.

Chapter 3 explores the notion of slowing down and outlines the role of tension, suggesting that cultivating the capacity to persevere through ambiguity offers new insights. This is followed by an outline which takes the reader through the architecture of the course synthesizing the series of encounters students moved through during the semester. This allows the reader to put into context the vignettes in Chapters 4–7.

Chapter 4 is the first of the student vignettes. This chapter focuses on a first semester art education doctoral student's tensions with a/r/tographical identities. The narrative follows Lauren's journey throughout the semester as she grapples with disrupting perceptions of self and the impact those perceptions have on research. Diving deep into the process of visualizing her ontology, epistemology, and axiological dispositions, she uncovers a complex understanding of how she knows her world expanding notions of research.

Chapter 5 explores the visual awakening of TWK, a doctoral Social Studies student as she interrogates her research questions. In surrendering and committing to the artist identity, her visual voice revealed its power with which to think. The emergence of her stick people uncovered an intraactive contemplation, conceptualization, and profound conversation with emergent knowing and understanding as she grappled with theory and the trajectory of her own research interests.

Chapter 6 explores the experiences of Brooke, a formally trained artist now pursuing a PhD in Art Education. Her perceptions and beliefs about the notion of (re)presentation are examined in-depth through her blog writings. The focus on tensions and a-ha moments in response to the readings are explored with an emphasis on play, bringing her attention back to process.

Chapter 7 explores how the experiences from the artistic research course are put into action with one student the following semester during an independent study. Through blog reflections and extensive discussion, the narrative explores the development of Gina's dissertation proposal while she simultaneously confronts the uncertainties and validity of her visual voice as research. By remaining in a state of curiosity and questioning, Gina comes to claim her way of knowing as a powerful method for research.

Chapter 8, while short, crescendos with additional questions, challenges, and invitations moving forward.

While there is a scholarly undertone to the writing here, it is my hope to make this accessible to all readers, while also being digestible and practical. It is for artists, students, thinkers, makers, researchers, and anyone just wanting to think/create beyond the margins. This is NOT about making art, it is not a method, or a prescribed way of doing things. It is an invitation to rethink the power of the visual, a bit on steroids, if you will. Visual literacy is powerful and deserves equal standing with verbal literacies. There is mounting evidence of the use of visual tools to make sense of the world and the movement in visual note taking is another crack in the wall of verbal privilege. Terms like "sketch note" (Rohde, 2013), image think, even doodling have been present for some time although much of that research centers on memory and retention. My own early doodling was evidence of that. The vignettes shared here illuminate emergent epistemologies through the lens of conscious ontologies, bringing new knowing into the world. This process is generative, creative, uncomfortable, multidisciplinary, and, and, and . . .

References

Felix, G., & Guattari, D. (1987). *A thousand plateaus: Capitalism and schizophrenia*. Trans. by Massumi, B.)., University of Minnesota, Minneapolis.

Ingold, T. (2010). The textility of making. *Cambridge Journal of Economics, 34*(1), 91–102.

Irwin, R. L. (2004). A/r/tography: A metonymic métissage. *A/r/tography: Rendering Self Through Arts-Based Living Inquiry*, 27–38.

Jackson, A. Y., & Mazzei, L. A. (2016). Thinking with an agentic assemblage in posthuman inquiry. In *Posthuman research practices in education* (pp. 93–107). Palgrave Macmillan.

Koro-Ljungberg, M. (2015). *Reconceptualizing qualitative research: Methodologies without methodology*. Sage Publications.

Manning, E., & Massumi, B. (2014). *Thought in the act: Passages in the ecology of experience*. University of Minnesota Press.

Patel, L. (2016). *Decolonizing educational research: From ownership to answerability*. Routledge.

Rohde, M. (2013). *The sketchnote handbook: The illustrated guide to visual note taking*. Peachpit Press.

Ruopp, A. (2019). Portrait of an a/r/tographer: Theory as conceptual medium. *International Journal of Education & the Arts*, *20*(9).

Ruopp, A., & Unrath, K. (2019). Making artistic learning visible: Theory building through a/r/tographical exploration. *Visual Arts Research*, *45*(2), 29–48.

Walrath, D. (2017). *Arts-based research in education: Foundations for practice* (M. Cahnmann-Taylor & R. Siegesmund, Eds.). Routledge.

2 A/R/Tography, Identity, and Theory as a Palette of Verbs

A/R/Tography as Conceptual Architecture

As noted in Chapter 1, the considered course was populated with students from varying disciplinary backgrounds with widely different research foci. As such, the architecture for inquiry in the course needed to be open and accessible for a multitude of interests while also being productively disruptive to already learned conceptualizations of research. Emphasizing the idea of learning and practicing through the multiple lenses of the artist, researcher, and teacher aligned with the tenets of a/r/tography provided a conceptual framework for putting theory into practice.

Irwin and Springgay (2008) describe A/r/tography as "a research methodology that intentionally unsettles perception and complicates understandings, [as such], we come to know and live within a space and time [that] is subsequently altered" (p. xxvi). Furthermore, they speak to notions of openings or openness as spaces and possibilities for conversation and the building of relationships. The described research process is active and relational and not passive or perceived from afar. Hence, each encounter in the course was designed to open spaces for experiential exchanges with each other as well as with objects, materials, and multimodal literacies. By inviting students into active inquiry, the lenses of the artist/researcher/teacher are put to work in productive ways; introducing the tensions of the unknown while simultaneously opening spaces for new discovery. The core course experiences were designed around the practices of a/r/tography with additional theoretical concepts encountered through inter/intra-action with material woven in. These encounters and explorations happened in the context of the classroom, building a community of a/r/tographical explorers as they discovered how materiality can organically emerge to express and reveal an idea. In real time and through authentic personal experience, they were able to see how ideas literally matter and how intention and attention shape them. According to Irwin (2004), "to be engaged in the practice of a/r/tography

DOI: 10.4324/9780367823818-2

means to inquire in the world through an ongoing process of art making and writing not separate or illustrative of each other but *interconnected and woven through each other* [emphasis added] to create additional and/ or enhanced meanings" (Sinner, 2017, p. 40). Historically, most traditional research methodologies privileged verbal texts, but this course offers an excellent platform to engage with inquiry differently from the beginning. A/r/tography as a fluid framework opened spaces for sensuous discovery. It invited inquiry into a process of finding rather than an emphasis on solving. In a/r/tography, "all three ways of understanding experience- theoria, practice, and poesis- are folded together and form rhizomatic ways of experiencing the world" (Irwin & Springgay, 2008, p. xxix). Thus, the complexities of active practice within the framework of a/r/tography are lived through the entwined identities of the artist/researcher/teacher.

Identity

In order for the students to fully understand the power of a/r/tography as an architectural mechanism for inquiry, I needed to frame an introductory experience that not only explored the identities as comingling voices offering contiguous yet discreet lenses but also frame "a/r/tography as a methodology of situations" (Irwin et al., 2006, p. 72). Building on concepts of the rhizome (Deleuze & Guattari, 1987), I imagined that the nuances of the artist, researcher, and teacher identities might be discovered and illuminated through the practice of visually exploring the actions of each. But how? I was beginning to develop my own working theory which posits that putting the action of theory into visual practice through multiple lenses complicates previously learned habitual modes of thinking and opens up spaces for new understanding by tolerating and moving through tensions and ambiguity. I wondered, what might happen when this unfolds in the context of a dynamic group with diverse backgrounds and knowledge bases?

Research and theory are sometimes intimidating words. Likewise, the identity of "artist" feels elusive for most. These are things that other people do; people in academe write scholarly papers while artists are tucked away in studios magically wielding brushes and producing powerful imagery. I would like to offer a more accessible conceptualization. In terms of research, we live the actions, research, and theory constantly. Perhaps, a less-complex version than formal research, nevertheless, the practice of discovery is all around us as we move through the tasks of life. We tinker, experiment, and solve problems every day. Likewise, we theorize continuously. Simply put, theory explains how or why something is. When put into material practice, theory is active and productive, inviting more sensual and concrete experiences which allow the learner to make sense of an otherwise philosophical or conceptual idea. In other words, the action becomes a mechanism for understanding. Artists explore materials and make critical and creative decisions about color, shape, and line texture to illuminate an idea or concept; they traffic in visual literacy. In an everyday context, the average person makes multiple creative decisions each day—from choosing an outfit to making dinner from what is in the pantry. These are creative choices that produce something in the context of lived experience. Making these concepts accessible and relatable to students was important. They actively make choices about attention and action in every moment.

Theory as a Conceptual Palette of Theoretical Verbs

This notion of action inspired wonder. How might this map on/in/to the active doings of research and a/r/tography? I quickly found myself listing performative actions and verbs that came to mind from my own experiences with a/r/tography: sensing, substituting, intertwining, intersecting, playing, relating, realizing, moving, opening, connecting, unknowing, negotiating, reverberating, and coexisting (to name a few). These were broad and needed more context. How did the specificity of each identity impact the action and intention of the verb? As I conceptualized the individual identities of the artist, researcher, and teacher, I realized multiple verbs could be associated with two or even three of the identities. The expression and intention of the verb, however, became different. This is where a deeper contemplation of the nuances of action pushed up against prior notions or limited definitions of ideas and words. How might using the visual expand and open space for a different kind of dialogue? What kind of experience or encounter might afford this open-ended space for discovery? How does our attention shape context and action?

In our first class, I offered my own experiences and understandings of a/r/tography by sharing my own artist/researcher/teacher experiences. This led to a discussion about what each of these identities is doing. Who and what is the identity in service of? What does the action of each identity produce? Why is conscious attention to this important? After sharing my own experiences and perspectives, I asked the students to conceptualize themselves as artists, researchers, and teachers. This was interesting in and of itself as there was a fair amount of resistance to at least one identity depending on disciplinary backgrounds. I asked them to consider what those particular identities represented in the world for them. What assumptions did they carry? Were they inherited? Was this identity self-proclaimed? Assigned? Cultural? What did each of these roles do or produce? I requested they translate these thoughts into actions, creating a list of verbs which they populated on the ***palette of verbs*** worksheet. While still a verbal exercise, the worksheet offered a visually mapped representation of potential intersections and

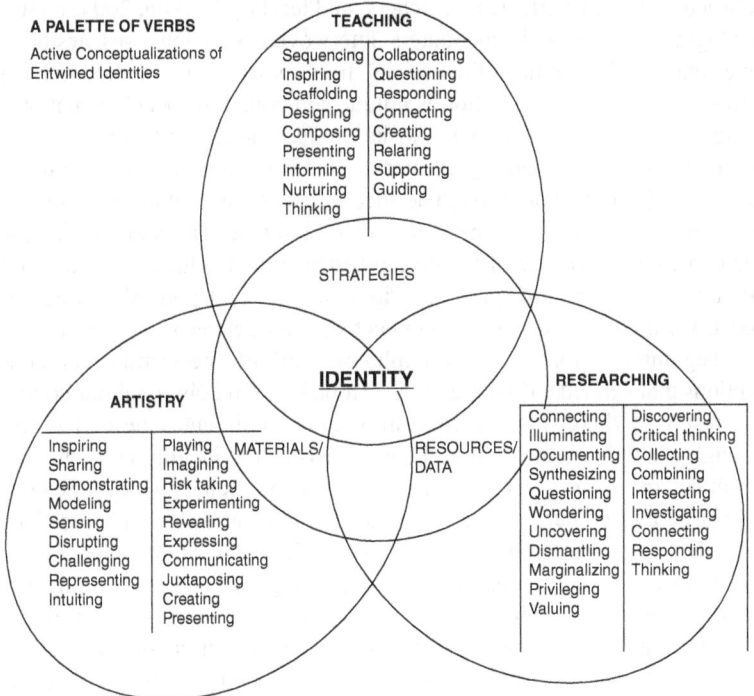

Figure 2.1 A palette of verbs

overlaps capturing the actions within the context of a specific identity while simultaneously untangling the nuances of difference.

The actions and doings of identity embody distinctly unique expressions relative to context. Exploring these nuances required an interrogation of attention. What are we really doing? Why? What influences it? How? What biases are we predisposed to and how might they impact what we think we see, know, and do in research? After individually exploring the actions of each identity utilizing the palette metaphor, we worked collectively. On a 10 feet × 4 feet sheet of paper, the students populated each identity with the actions they had recorded on their own worksheet. Students moved rhythmically around the large paper adding their verbs around each identity. As everyone moved around the paper adding their verbs, spontaneous conversations broke out, making comparisons, noticing like verbs under different identities, and speculating what they meant. The central focus of the three identities became secondary to what was emerging in between and all around, "it is not about dichotomous thinking but rather dialogical thinking, relating, and perceiving. It is about living in the borderlands, the spaces between and amidst artists, researchers, and teachers" (Irwin, 2004, p. 30). A/r/tographical research encounters interwoven with sensuous aesthetic investigation differentiate themselves from other more methodological forms in that each investigation is unique and bound to specific situations, context, time, and is continuous in its unique nature. Furthermore, practicing "theory as a/r/tography creates an imaginative turn by theorizing or explaining phenomenon through aesthetic experiences that integrate knowing, doing, and making; experiences that simultaneously value technique and content through acts of inquiry; experiences that value complexity and difference within a third space" (Irwin, 2004, p. 31). Upon exhausting our verbs, we then started drawing connective lines between like verbs. This was beginning to form a very complex network of intersections and connections that echoed a rhizome. As we stood back and observed our collective response, there was a realization that one might enter these identities through multiple points. It was the emphasis of the identity and lens that might determine the specific intention of an action. Being aware of the lens, emphasis, and its purpose was important in making conceptual decisions around creating, researching, or teaching. This was our first collaborative, community building a/r/tographical undertaking. We had created our first visual/verbal artifact which could be read and responded too in multiple ways and catalyzed a very rich platform for authentic discussion and reflection. As a group, there was a palpable sense of being on the precipice of something new, something different.

This exercise embodied and performed Irwin's (2008) notion of the singular plural.

> We are singular plural beings that are part of the whole of being singular plural. This is significant for a/r/tographers as they understand the need to be engaged in their own personal pursuits while they are contiguously positioned alongside the pursuits of others, and together are becoming a whole constellation of pursuits.
>
> (p. 72)

Because this occurred during the first class, many students were taking risks and allowing themselves to be vulnerable in their participation. Nevertheless, in response to the previous quote, one student (highlighted in Chapter 4) stated:

> *This made my brain hurt. I understand the complexity of this . . . I think. The reality of the messiness in this felt like way too many people had shown up at the dinner table and I didn't have room for everybody. I am*

Figure 2.2 Students exploring the intersections of a/r/tographical identity

Figure 2.3 Students noting similar actions of identities while discussing how each
lens has a nuanced difference in action and intention

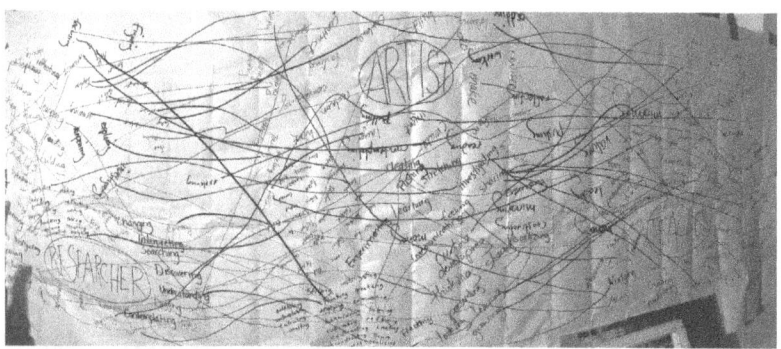

Figure 2.4 Final a/r/tographical map, rhizomatic in nature

> *such an advocate for sharing your craft and learning from others. I*
> *also think it is so important that you not only consume content from*
> *others, you also put content out in the world for others. I felt like this*
> *was an instance of the rhizome which felt really chaotic for me. I some-*
> *times feel like I differ from other creative types in that I work very*
> *methodically and like to work where it's quiet and solitary. Hopefully,*
> *I will be able to push through the discomfort in the chaos of creating*
> *around others and engage in an a/r/tographic community of practice.*
>
> *(Blog entry, LH)*

A/r/tography has continuously evolved since its initial inception in 2003 and has taken on many shapes, actions, and forms across different disciplines (Sinner, 2017). I imagine this as a complex ripple outwards embracing possibility and blurring the boundaries of disciplinary silos. Sinners' recent comprehensive review of a/r/tography seeks, among other things, "to situate a/r/tography as a form of inquiry" (2017, p. 40). The diversity of practice emerging from a/r/tography further confirmed this would serve my students well. A/r/tography's connection to poststructural theories makes it an excellent transformative vessel for exploring the lived engagement of theory and how it matters with/in research regardless of disciplinary backgrounds. In addition, the nomadic spaces of a/r/tography offers multiple entry points for engagement, exploration, experimentation, and play. This was affirmed by our initial group exercise. Especially important is the notion that "a/r/tography is a form of representation that privileges both text and image as they meet within moments of métissage" (Irwin, 2004, p. 35).

Imagine a conceptual playground, rhizomatic in nature affording multiple intersections, hyphenated relationships, and spaces. All are opportunities for identities to comingle and literacies to intersect. What possibilities *matter* when we play?

Practicing-Playing-Imagining-Visualizing

With this more comprehensive understanding of a/r/tography, let us try on these lenses and explore the conceptualizations of theory in action. In the next section, I return to my own experiences grappling with and making sense of a number of poststructural and post-human theoretical concepts. This artful reimagining is not inclusive of every perspective nor is it meant to represent an in-depth interrogation of each theory, although I have explored many of them with rigor. For me, this serves as a disruption, a questioning, and an experimentation with thought and action and how they blend with the visual to make meaning. I offer my own personal abbreviated conceptual vignettes, as a way of sharing my own positionality within this space. I share this here as it offers a glimpse into the concepts animating the theories that excite me. It is just one of the infinite number of ways to explore and play with literacies. Sinking into this process disrupted any fixed notions I held around my artist materials and also afforded me an opportunity to experience what I would later ask of my students.

As an artist and visual thinker, when I need to do a lot of writing, I have to think with metaphors and images. As I noted in Chapter 1, it is through this intra-action of literacies that I can then articulate with more clarity my discoveries and learning and apply them to research and/or curriculum design. My own personal investigation with my palette of verbs and their connections to the performative aspects of theory became a metaphorical journey which further informed my pedagogical thinking about what was unfolding in the class as I worked in my studio.

Pause here for a moment and visualize. Bring to mind your image of an artist in action. Imagine the artist drawing. How does it feel? What does it sound like? There is much more going on than an arm moving as it guides the hand which holds the pencil which moves across the paper to develop an image. Go deeper; visualize a symphonic orchestration of knowing/feeling/being arranged in different patterns, breathing with slow and quick rhythms and how that *being and feeling* impacts knowing while acting and visualizing. How does an image emerge? Is this skill? Inspiration? Is it human centered? Material driven? How does emergent knowledge acquisition unfold when all the senses are engaged? "For the multi-sensory researcher—as a seeing, perceiving, knowing organism—the sensory, affective, and empathetic dimensions of being human are central to the research process" (Pink, 2011, p. 606) When visually engaged, there is a complex orchestration of actions which emerge in a nonlinear tangle of mattering. In my experience, an artist might be actively re/searching both internal and external stimuli, concepts, memories, inspiration, and skills in a quest to express and discover newness. For me, thinking and doing from this multi-sensory theoretical perspective, I hope to expand on possibility.

Cotter (2019) speaks to an important aspect of artistic research which plugs into the mechanisms of a/r/tography, offering that artistic research "follows its own inner logic in residence with the wider languages and sensibilities of art, rather than the logic of any external discipline" (p. 13). The inner cultivation of one's unique creative process (artist or not) opens a space of limitless possibility and the "associative logic of artistic thinking" (p. 13). My own experiences with a/r/tography afforded me opportunities to explore various, sometimes contradictory theories as a deliberate disruptive action. While I recognize that theories have historical lineages which arise from varying ontological dispositions, I was interested in what the action of a theory might produce looking forward. What new questions might arise? What might be unsettled by blending seemingly contrasting ideologies? As I explored what might happen, I began to visualize/conceptualize theories as actants, abstract actions, tools to be blended, layered . . . assembled and disassembled from my artist lens. I began by connecting a verb or two with a theory and then blended visual attributes into the mix as a tool for

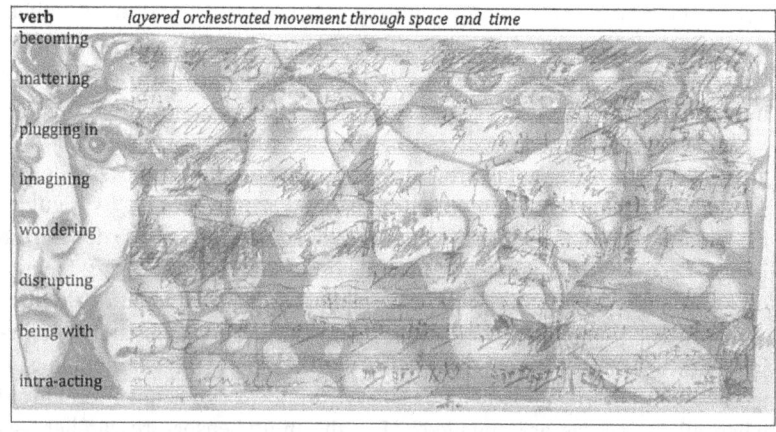

Figure 2.5 A/r/tographical communities; a symphonic mapping of intersecting verbs

reconceptualizing and expanding upon the action as something manifest or visible beyond the word or action.

Much like the pallet of verbs described earlier, I expanded the pallet to encompass theory as well. I experimented and played with this conceptualization exploring theory as visual expressions and what might happen as I think with theory through visual expressions blurring, layering, assembling, and intra-acting. As I explored a theory, I associated it with an action/verb. The verb was contextually relevant but certainly not the only possibility; it is simply the one that arose as I surrendered to the feeling within my own understanding of each theory. This was then explored through different materials, until one emerged as a partner in my contemplations. The material acted as both an expression and a metaphor.

Amy's Conceptual Palette of Sensory Play

Theory: Posthumanism
Verb: being with
Visual Manifestation: Conceptual color: Titanium White (tints)

One element of posthumanism, as described by Braidotti (2013), emphasizes the notion of post-anthropocentrism, decentering man as the "the measure of all things" (p. 67). In the decentering of man, language is then also displaced as the centrally accepted mode of communication, opening the ontological door for intersubjective intra-actions between animal, Earth, and organic others. Classical parameters, which previously regulated communications, now disrupt language, as it is no longer the privileged signifying system of engagement, and this opens the door to human and non-human exchange. Rooted in an ethic of joy, posthumanism calls upon visionary thinkers to imagine conceptually creative forms of intersubjective relationality. Braidotti concedes that these "qualities . . . are neither especially in fashion in academic circles, nor highly valued scientifically in these times of coercive pursuit of globalized excellence" (2013, p. 191). However, I suggest that they are critical as they illuminate new ways of understanding required for interconnected, digital, and globalized thinking. Translated into an artistic conceptual medium, posthumanist thought decenters the artist "I" as the creator.

The artist becomes a nuanced medium in the service of emergent knowledge. Tints expressed as both a concept and a material, illuminate, transform, or hide meaning and are empowered by their partner-light. The artist participates as one instrument with/in an elegantly collaborative collective. Illuminating the margins, shining, or casting shadows through the material (pens, paper, and paint) and theory, meaning ripples across surfaces. The

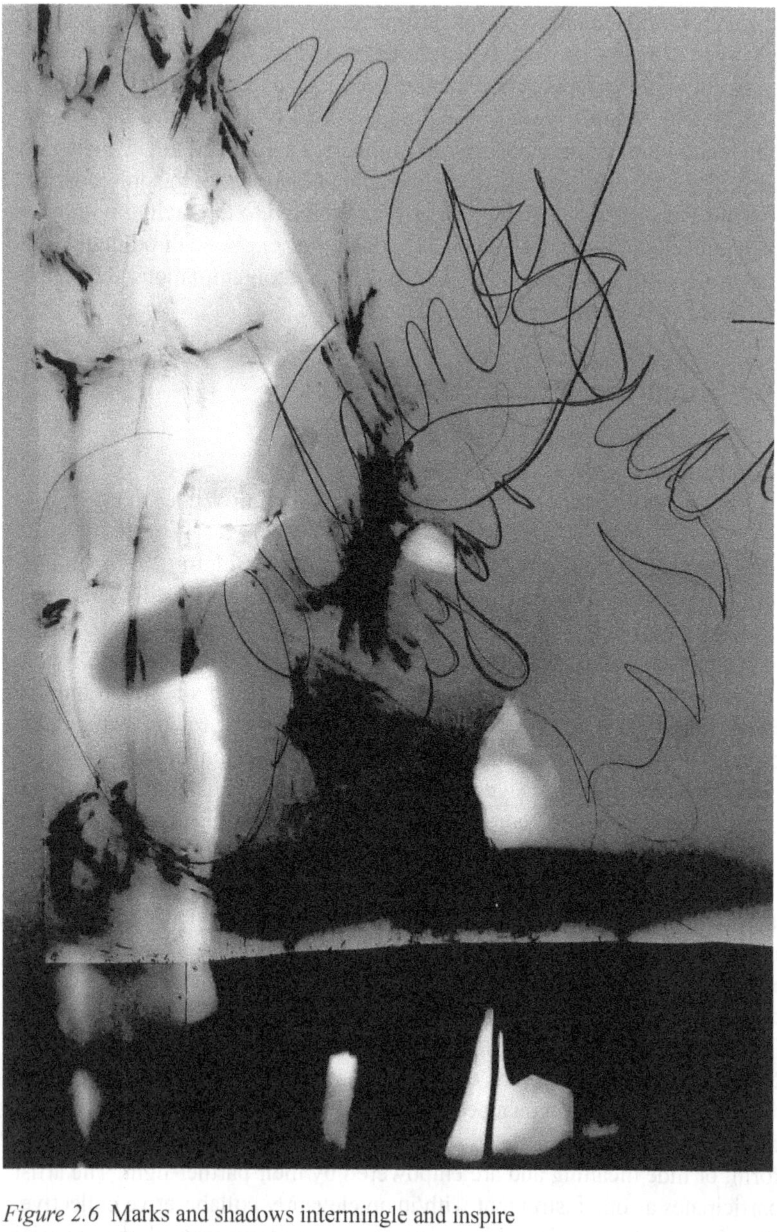

Figure 2.6 Marks and shadows intermingle and inspire

process expresses agency in these emergent moments, which animates the artist, material, and concept.

Theory: Agential Realism
Verb: intra-action, blurring
Visual Manifestation: Conceptual tools; Walls, scissors, bushes, paper etc.

Interconnected with posthumanism, in agential realism, matter articulates itself as an ontological performance of the world. Barad (2007) believes "humans are neither pure cause nor pure effect but part of the world in its open-ended becoming" (p. 148). Barad goes on to speak of matter as becoming. It arises and materializes through intra-action. Material expresses agency. "Agency is not held, it is not a property of persons or things; rather, agency is an enactment, a matter of possibilities for reconfiguring entanglements" (Barad, 2012, p. 55). Barad goes on to explain that agency requires attention to *response-ability*. In the studio, I am intra-actively engaging in the phenomena of emerging theory. It becomes visible as emergent matter through artistic encounters. Thus, the artist, movement, media, paper, iPad, video, studio space, and theory materialize through intra-active responses; therefore, "matter and meaning are mutually articulated" (Barad, 2007, p. 152) within and through each other.

For me, as an artist who operates from a posthumanist paradigm, there is no ontological separation between the artist, the action of doing, the material, or the many spaces of realization that unfold and enfold theory. Binaries are erased while ambiguity and tension remain, providing a threshold for emergent discovery. All things are *doing*, engaged in constant intra-action, and therefore, for me, my way of knowing and being is ontoepistomological (see Table 2.1).

Theory: Diffraction
Verb: breaking, emerging, and mattering
Visual Manifestation: Conceptual emphasis; process

Diffraction transforms the original rather than reflecting back what is already known. Within this, the intra-actions of art materials, theory, space, and the action of the emerging image, knowing are transformed. Barad describes diffraction as entangled phenomena. Layers of habitual thinking are peeled away in action, as material is tangled with time, space, movement, and mattering. Barad (2012) points out that "diffractive readings bring inventive provocations; they are good to think with" (p. 50). The imagery and creative performance of thinking with theory in the studio space therefore diffracts traditional methods of reading and doing research. Table 2.1 shows Barad's

Table 2.1 Artistry as diffractive practice

Diffraction as quoted from Barad (Barad, 2007, pp. 89–90)	Amy's Artistic Process
diffraction pattern marking differences from within and as part of an entangled state	**collision** past meets present, entangled temporality, moments of transcending habit/stuckness
differences and relationalities objectivity is about taking account of marks on bodies, that is the differences materialized, the differences matter	**invisible to visible** objectivity is about response/able conceptual weaving into matter
diffractive methodology performativity subject and object do not preexist as such but emerge through intra-actions	**expressive tension performativity** subject(idea) and object(material) do not preexist as such, but emerge through intra-actions
entangled ontology material-discursive phenomena	**entangled a/r/tography** blurred blended emergent identities
onto-epistem-ology knowing is material practice of engagement as part of the world in its differential becoming	**a/r/t-onto-epistem-ology** knowing is visual material practice of intra-engagement as part of the world in its differential becoming
intra-acting within and as part of differences emerge with phenomena agential separability; real material differences but without absolute separation	**intra-acting within and as part of differences emerge with and through phenomena and between spaces** agential realism; material and conceptual differences become invisible intersections
diffraction/difference pattern intra-acting entangled states of nature cultures	**diffraction/difference pattern** intra-acting entangled states of materials and literacies, words become visual, agentic imagery
about making a difference in the world about taking responsibility for the fact that our practices matter; the world is materialized differently through different practices (contingent ontology)	**about making a difference in the world of knowing and understanding** about taking responsibility for the fact that our practices matter; the knowledge is materialized differently through different practices (contingent ontology)
phenomena are objective referents accountability to marks on bodies, accountability and responsibility, taking account of differences that matter	**active phenomena are not fixed/replicable** multidimensional, multimodal and hypermodel accountability to process of emergence honoring differences that matter

Diffraction as quoted from Barad (Barad, 2007, pp. 89–90)	*Amy's Artistic Process*
ethico-onto-epistem-ology ethics, ontology, and epistemology not separable	**ethico-onto-epistem-ology** ethics, ontology, and epistemology not separable
reading through (the diffraction grating) trans-disciplinary engagement attends to the fact that boundary production between disciplines is itself a material discursive practice; how do these practices matter? subject, object contingent, not fixed respectful engagement that attends to detailed patterns of thinking; each fine-grained detail matters	**reading through visual/conceptual (the diffraction grating)** multimodal engagement disrupts habitual ways of encountering knowledge blurring boundaries between research practices, which privilege material discursive practice; how does the blending of multimodal texts and conceptual theory matter? subject, object emergent, rhizomatic, and feeling intuitive becoming fluid respectful engagement that attends to detailed patterns of thinking; each fine-grained detail matters
Summary accounting for how practices matter	**Summary** accounting for how blending literacies and conceptual practices matter

explanations of diffraction (on the left). Removing Barad's comparison with reflection, I have instead offered my own artist's interpretation and application emphasizing my process in the corresponding right-hand column.

> *Theory:* Assemblage
> *Verb:* plugging-in
> *Visual Manifestation:* Conceptual medium, collage, paint, and glue

If we reference a dictionary, assemblage seems very straightforward. As a noun, it implies a *thing*. However, as a theory, the word is transformed into an action or verb. Jackson and Mazzei (2011) suggest, "it is the process of making and unmaking a thing" (p. 13). They go on to offer that these *things* include conceptual notions of thinking through data with theory and thinking about data as theory. Plugging data into theory and theory into data eliminate a binary where one thing is doing something to something else; rather, it becomes *with* the other, revealing something different.

My conceptual artistic translation and the doing and undoing of an assemblage is the merging of theory as an active and agential conceptual medium expressed through matter. A collage comes to mind but the collage itself is the manifest product of highly complex visual engagement with image and material and expresses only one of the many number of possible outcomes. Assemblage then is an iterative process. The process is the nondefinitive product revealing layers of insight which otherwise might remain hidden. An assemblage offers infinite possibility (Whitehead, 1968), which mirrors artistic process. Realizations through encounters with materials, ideas, feeling, memory, intention, etc. inform action. As discussed briefly previously, the studio, the paper, and the material, all have agency and are intra-active within the process. As an artist, I wander from drawing to text to drawing to writing and back to the drawing again. Knowing is assembled through the conceptual assembling of multimodal texts.

Theory: Rhizome
Verb: mapping, becoming, and tangling
Visual Manifestation: Conceptual medium and line/texture

To try and define the rhizome as Deleuze and Guattari speak to it contradicts the very notion of what it is—indefinable. It has no beginning, no end, and seduces with infinite possibility. Deleuze and Guattari's writing wanders across the page and embodies the ideas it speaks to, leaving me to find my own meaning. In some ways, what they have done is to create the textual equivalent of the mostly unspoken creative process. My own reading of their text and others like it takes on visual mappings as I link concepts and ideas, allowing my own encounter with the words to emerge without preconceived notions of meaning. Deleuze and Guattari's rhizome embodies my own ideation process. It closely echoes my creative process and may explain why I am so drawn to their theories. They feel inexplicable in purely text form. Potential lines of theoretical flight within and from the rhizome include; becoming, unfolding, rupturing, emerging, intensities, strata, and desire, all of which make frequent appearances in my artistic process. Their writing almost begs a visual intra-action, teasing the engagement of engaging in a different kind of response.

Theory: Genealogy
Verb: interconnecting, troubling, and transcending
Visual Manifestation: Conceptual color: gray

My own historical relationship with knowing is rooted in textual privileging. This human bias towards text (and do not get me started on numbers) has suppressed my visual voice for a long time. In this, my artist self seeks to disrupt the binary between text-based and visual-based knowing through the troubling and diffracting of genealogical inherited biases. "Foucault describes genealogy using one of Nietzsche's well-known metaphors.

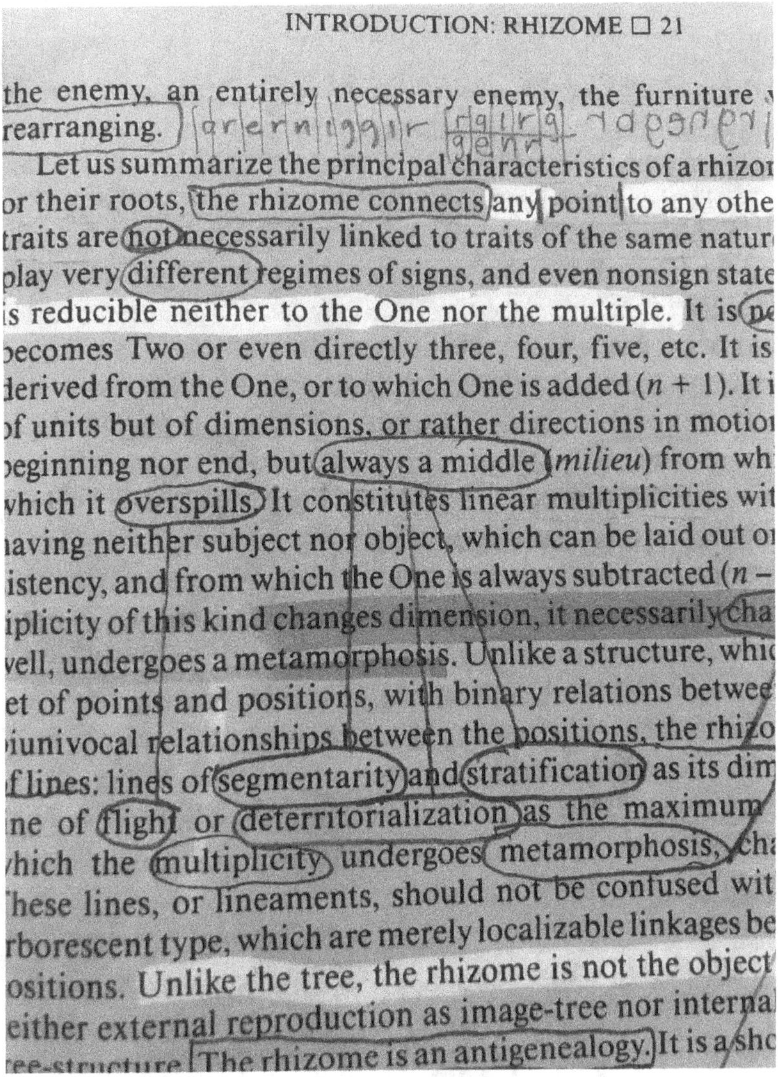

INTRODUCTION: RHIZOME □ 21

the enemy, an entirely necessary enemy, the furniture
rearranging.
Let us summarize the principal characteristics of a rhizo
or their roots, the rhizome connects any point to any othe
traits are not necessarily linked to traits of the same natur
play very different regimes of signs, and even nonsign state
is reducible neither to the One nor the multiple. It is n
becomes Two or even directly three, four, five, etc. It is
derived from the One, or to which One is added $(n + 1)$. It i
of units but of dimensions, or rather directions in motio
beginning nor end, but always a middle (*milieu*) from wh
which it overspills. It constitutes linear multiplicities wit
having neither subject nor object, which can be laid out o
istency, and from which the One is always subtracted $(n -$
iplicity of this kind changes dimension, it necessarily cha
vell, undergoes a metamorphosis. Unlike a structure, whic
et of points and positions, with binary relations betwee
iunivocal relationships between the positions, the rhizo
f lines: lines of segmentarity and stratification as its dim
ne of flight or deterritorialization as the maximum
which the multiplicity undergoes metamorphosis, cha
these lines, or lineaments, should not be confused wit
rborescent type, which are merely localizable linkages be
ositions. Unlike the tree, the rhizome is not the object
either external reproduction as image-tree nor interna
ree-structure. The rhizome is an antigenealogy. It is a sho

Figure 2.7 Reading differently

Genealogy is "gray", its task being to *decipher* the hieroglyphic script of humans' past, a past that is neither black (i.e. totally unknown) nor white (i.e. transparent), but something in between (gray), that is, *ambiguous* and *uncertain*" (Sembou, 2011, p. 3). The ambiguous and uncertain dwell in the spaces in between what was and what is.

Figure 2.8 The addition of marks slows reding down makes a space for moving through tensions of not knowing

As an artist, this requires acute attention to my reflexive process as I create. Noticing my own subjective privileging of certain marks, materials, and recurring images, which keep me stuck in a comfortable knowing, jars how I make sense of meaning, whether data for research or personal expression. Breaking, blurring, and confronting my own inner esthetic and sacrificing it to an unknown yet to emerge requires great trust. Hence, my high-contrast black and white drawings moved into grays and finally were disrupted with the addition of color, and yes . . . text.

Finally

The role of A/R/Tography is a dynamic and fluid framework for open-ended exploration. The elegance of a/r/tography, while rigorous, also offers unique platforms for investigation honoring the unique lived experiences of each artist/teacher/researcher. As a creative framework, it offers multiple entry points into experience. The slash ("/") in-*between* each letter becomes a nomadic space or a threshold for lines of flight, diffraction, disruption, and the creative unfolding of new knowledge. For me, it is (in)between each identity and literacy, where theory is woven and performed. In assembling, mapping, and layering the a/r/tographical identities of the artist, researcher, and teacher, the nature of the rhizomatic structure is "open and connectible in all of its dimensions; it is detachable, reversible, susceptible to constant modification" (Deleuze & Guattari, 1987, p. 12). Each identity commingles and inter/intra-acts with data and theory differently.

Artistic thinking, creative production, and inter and intra-action with materiality does something different. One does not have to self-identify as an artist to participate in this sort of thinking. One only has to surrender and trust in the process and allow the emergence of new knowledge to unfold within the relational experiences with the material. Receptivity to ambiguity, the disrupting of language as it becomes entangled with image, material intra-actions, and the tensions these moments produce offer a reconsideration as to how research unfolds.

References

Barad, K. (2007). *Meeting the universe halfway: Quantum physics and the entanglement of matter and meaning.* Duke University Press.

Barad, K. (2012). Interview with Karen Barad. *New Materialism: Interviews & Cartographies,* 48–70.

Braidotti, R. (2013). *The posthuman* (p. 26). Polity.

Cotter, L. (Ed.). (2019). *Reclaiming artistic research.* Hatje Cantz.

Deleuze, G., & Guattari, F. (1987). *Capitalism and schizophrenia: A thousand plateaus* (B. Massumi, Trans.). University of Minnesota Press.

Irwin, R. L. (2008). Communities of a/r/tographic practice. In *Being with a/r/tography* (pp. 71–80). Sense Publishers.

Irwin, R. L., Beer, R., Springgay, S., Grauer, K., Xiong, G., & Bickel, B. (2006). The rhizomatic relations of a/r/tography. *Studies in Art Education, 48*(1), 70–88.

Irwin, R. L., & De Cosson, A. (Eds.). (2004). *A/r/tography: Rendering self through arts-based living inquiry*. Pacific Educational Press.

Irwin, R. L., & Springgay, S. (2008). A/r/tography as practice-based research. In S. Springgay, R. L. Irwin, C. Leggo, & P. Gouzouasis (Eds.), *Being with a/r/tography* (pp. xix–xxxiii). Sense.

Jackson, A. Y., & Mazzei, L. A. (2011). *Thinking with theory in qualitative research: Viewing data across multiple perspectives*. Routledge.

Pink, S. (2011). *The Sage handbook of visual research methods* (E. Margolis & L. Pauwels, Eds., pp. 600–626). Sage Publications.

Sembou, E. (2011, June). *Foucault's genealogy* (pp. 16–17). 10th Annual Meeting of the International Social Theory Consortium, University College Cork.

Sinner, A. (2017). Cultivating researchful dispositions: A review of a/r/tographic scholarship. *Journal of Visual Art Practice, 16*(1), 39–60.

Whitehead, A. N. (1968). *Modes of thought* (Vol. 93521). Simon and Schuster.

3 Tensions

Reading Differently, Reading Diffractively

Now that we have established the conceptual framework animating the ideas in this book, let us move into the notions of the process, the nuts and bolts of action, and its relationship with attention and conceptions of time.

When I visit art museums, I am fascinated by the way people move through the space. I spend as much time observing them and their encounters with the artwork as I do with the artwork itself. In many cases, these museum goers are either in constant movement from piece to piece or, worse yet, pausing only minimally to take a picture or selfie with an artwork. I recall hearing once that a typical person walking through an art exhibit or museum "reading" the art work spends an average of 30 seconds with any given piece. That amount of time is less if the piece strays from their esthetic preference. Attention is typically drawn to the focal point or some aspect of the work that attracts attention or feels accessible in some manner. There is a brief moment of contemplation, possibly inspiration or connection, and then they move on. How can deep understanding or meaning be acquired and made sense of in such a short encounter? In these brief moments, the complexity of the entire composition may remain unseen. The details, the layers, and the edges where imagery is alive and speaking with the rest of the composition in a polyphonic chorus of color, line, shape, image, meaning, etc. are passively consumed without conscious thought or attention.

For many who encounter academic texts, a cursory linear read of scholarly writing may be consumed with more attention due to its consistent and more familiar linear structure. However, our eyes pass over words and sentences quickly, consuming words readily accessible to the reader with other parts relegated to the margins. Both visual and textual artifacts are the complex summation and intersections of multiple experiences and moments. Authors and artists alike accumulate experiences that are integrated into a finished piece. These artifacts are, as Manning and Masumi might say,

DOI: 10.4324/9780367823818-3

Visual Arts Research   Volume 45, Number 2   Winter 2019 | 29

Making Artistic Learning Visible:
Theory Building Through
A/r/tographical Exploration

Amy Ruopp
University of Missouri

Kathy Unrath
University of Missouri

I do not seek, I find.
—Picasso

If art educators are going to teach to meet the needs of the 21st-century learner and an ever-changing global society, art education needs to move beyond teaching content and embrace creativity, process, and ideation as interdependent components in teaching and learning art. Communicating and teaching the complexities of ideation and process are critical for reading and engaging an increasingly visual world. This article offers insights into how multimodal reflection in action and on action intersecting with rhizomatic thinking creates conditions for paradigm shifts about identity and conceptualizations of creative knowledge acquisition with pre-service art education students. By capturing the multimodal moments of their own creating through video format, a layered text capturing material, action, voice, intention, and reflection emerges to make process visible. Pre-service teachers entering the field of art education cognizant of their epistemology and the "how" of knowledge construction not only empower their students as active learners but are also living advocates for art education.

Keywords: Reflection, A/R/Tography, identity, creativity, rhizome

Inspirational Roots

This study began 3 years ago during an internship with my advisor in an art education methods course at a midwestern research-intensive university. It was inspired by conversations with pre-service art education students, graduate and undergraduate, who revealed to us surprising perceptions about their professional

Figure 3.1 Visual image juxtaposed with textual, both are read, but read differently

Figure 3.2

"co-composed" throughout time. Therefore, the complexities that created these artifacts live hidden within them.

What might unfold if viewers and readers practiced slowing down and sinking into a text, be it verbal or visual, and that slowing down invited a different, deeper read? If you spend 10 minutes in front of a painting, the encounter changes. Where do your eyes wander? What if a viewer sat for an hour? Two? What if we persevered through the ambiguity or tension of not understanding? A much richer and complex understanding begins to surface when *time* is invested in wondering and seeing more than the initial first pass. So, it is with reading dense academic theoretical text. Revisiting and wondering what something means from multiple perspectives can offer insights and enable readers to make connections which then translate into contextually relevant practice. Reading does not need to be passive in nature; it can be participatory. What if we inter/intra/acted with texts and we diffracted literacies by reading them through one another? "Diffractive readings bring inventive provocations; they are good to think with. They are respectful, detailed, ethical engagements" (Barad, 2009). As an a/r/tographer, I am inspired by the co-mingling of texts and a commitment to an investment in the time it might require to read deeply. As such, my own experiences reading verbal and visual texts through each other illuminated exciting ways of addressing my own research questions.

Time for (A)tension

Reading is a learned task. For most English-speaking cultures, we are taught early on that we start on the left top side page and we move from left to right, line by line, from top to bottom. There are rules to the writing, grammar, punctuation, sentence length, paragraphs, etc. This architecture is provided so that the words woven into sentences, which weave into stories and/or information, follow some manageable linear form of thought. It works. Just like most things, it is important to remember that at some point, somebody made this up. It was invented over time. Reading and writing, the way we typically encounter them, is, in essence, a sort of prescribed methodology. It serves a function and purpose, just like many of our tried-and-true research methodologies, and is similar to the seven elements of art and seven principles of design in art.

It has been my experience that students sometimes struggle to make sense of densely written theoretically rich research texts. In the first semester of my own doctoral program, I was assigned a reading in a Social Theory course. It was 14 pages and took me 10 hours to read because I thought I had to understand every word. I was literally in tears because I thought I had to make sense of the words, with more words. Words I had never seen.

I have observed this phenomenon in my own students as well! Tensions often arise from not knowing or feeling unable to immediately make sense of what they are reading. I discovered, however, that staying with these tensions, embracing ambiguity, and visually dialoguing with written text offered a different path into knowing, clarity, and connecting to the theory in meaningful ways. Offering encounters with dense reading by slowing down and sinking into particularly difficult parts of the writing through a multi-modal dialog with conceptual and physical matter affords opportunities for something different to occur. Inviting visual materials and processes into theoretical exploration further "decenters, troubles, and resists the human-ist subject and the [academic] structure that enables it" (St. Pierre, 2017, p. 1080). With(in) tension is the potential for growth. Cultivating the capac-ity to embrace tension as a possibility, rather than giving way to paraly-sis transforms how students engage with research. Through this practice, we disrupt traditional notions of engagement with reading, research, data, analysis, etc. Below are quotes from students' blogs that both illustrate and speak to this phenomenon. Their words are italicized while the bold text indicates a citation from the course reading that they are thinking with or through.

> *I seem to be able to "draw" or illustrate more quickly than words. It is different to take up this assignment first with images and then write text to construct or interpret my thinking place. Sousanis (2015) draws from Masaki Suwa and Barbara Tversky (2001) to address the process of drawing* **"is a means of orchestrating a conversation with yourself" (p. 79) and that "we draw not to transcribe ideas from our heads, but to generate them in search of greater understanding . . . distributing mental processes between "conception and perception" (p. 79).**
> *(TKW, 2018, blog entry)*

> *In my visual representation I literally drew a plug . . ."plugging in" to this idea in my head about what a visual representation of this assem-blage of connectives, and ambiguity in an effort to see what ideas emerge. The different parts that are doodled are symbolically the differ-ent ideas that are emerging from looking in and in between the spaces in different and unhinged ways. This process forces the researcher to see the spaces differently and see how all of the thinking, data, and theory work together in ways we have yet to fully understand*
> *(LH, 2017, personal blog entry).*

Many poststructural theories (diffraction, assemblage, rhizome, posthu-manism, etc.) echo the essence of creative acts carried out daily by artists

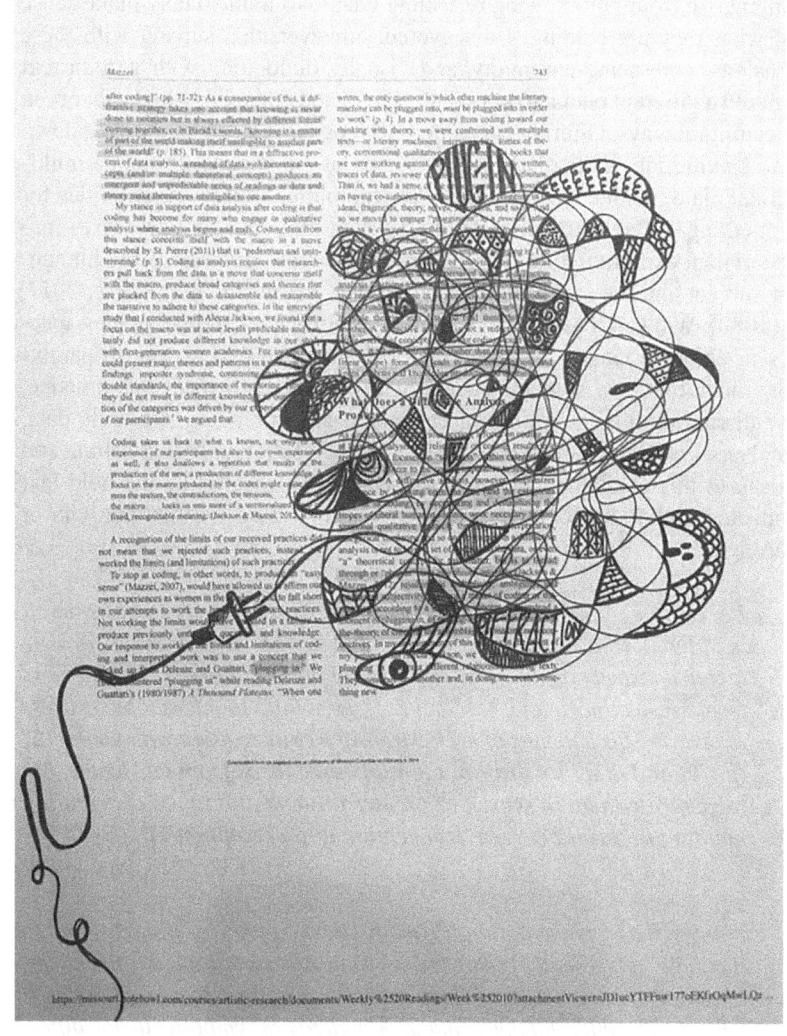

Figure 3.3 Collection of student visual intra-actions with text

in their studios. As an artist in conversations with other visual creatives, we all thrive in a space of wonder, curiosity, disruption, ambiguity, and tension so that concepts emerge organically without a fixed destination in mind. In this, new conceptualizations of an idea are expressed through materials to make meaning. For us, this is postqualitative visual research.

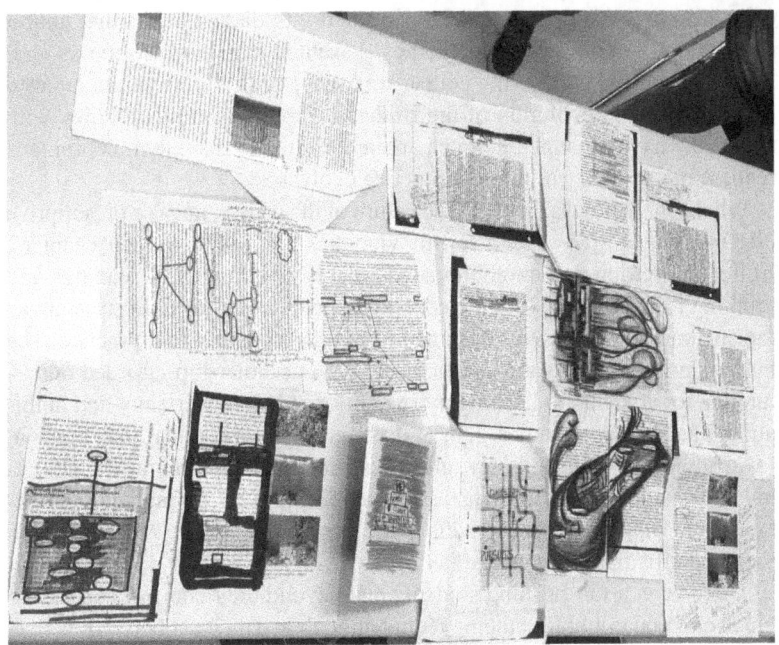

Figure 3.4 One student's visual interaction with text

It is rigorous, focused, and profound in what it uncovers. Research is about process. In contemplating qualitative research, Springgay and Zaliwaska (2015), quoting Manning and Massumi (2014), shift their focus to this notion of process over product/destination. In other words, positing the conditions or terms of research *before* the exploration or experimentation "results in stultifying its potential and relegating it to that which already fits within pre-existing schemata of knowledge" (Manning, 2014, p. 4). We must, Manning contends, "find ways of activating thought that is experienced rather than known, that is material and affective, and where experience accounts for 'more than human' encounters" (p. 136). That was a driving force within the artistic research course which I will outline next.

A Setting for Serious Play

As a teacher of teachers, I often ask students, what do you want your curriculum to do or produce? The same questions hold for art and research. What do you want it to do or produce? This begged the question, what did I want this artistic research course to do or produce? I wanted to create tension, incite disruption, tease out realization, cultivate transformation, offer affirmation and, and, and. . . I wanted it to disrupt habits that inhibit research and empower visual literacy. I wanted it to surprise me as well as my students. I believe the course was successful. Further, I am amazed by the continued evolution of the students who went on this journey with me. Many have remained in touch and are excited to share their continuous realizations as their multimodal practice evolves.

What I share next is the literal structure of the course, not as a prescriptive offering, but to give context for the vignettes that follow in later chapters. I offer the setting and an abbreviated weekly progression of readings and thinking to further flesh out the context, concepts, and the fluid structure of the encounters which were designed to put theory into visual practice. The integration of creative and artistic practices as scholarship-afforded opportunities for students to become more comfortable in uncertainty and ambiguity. I asked them to unlearn, or at least set aside, research as they knew it. I was inviting a playful messy unknown into the picture, literally. Cultivating a consciousness of the artistic and creative process through experimentation and play supports and refines the construction of new knowing and enriched articulated meaning-making.

There is an art to pushing students just beyond their comfort zones. Too much and you might lose them, yet not enough and they get bored or worse, comfortable. Therefore, as I designed each week, I felt a rhythm or a heartbeat was a good metaphor for how I introduced course readings and encounters. A really challenging week was almost always followed with something more easily accessible. In addition, the course was set in an art studio rather than a more traditional classroom, inviting a different encounter with learning spaces. As such, I share an abbreviated outline of the course here;

- Overarching conceptual structure

  - Readings (verbal)
  - Weekly Studio Investigations (visual)

- Each class was constructed around a:

  - Theme/Theory
  - Verb/Doing
  - Visual Research Medium
  - Artist

Week 1:

Thematic Big Idea/Conceptual Medium: *Identity/Rhizome*
Verb: *Perceiving*

Visual Research Medium of the week: *Mind mapping*
Exemplary Artist (a/r/tographers): *Kathy Unrath*
Readings: *A/R/Tography as Practice Based Research*; Irwin
Encountering Pedagogy through Relations Art Practices; Irwin and
  Donoghue

Week 2:

Thematic Big Idea/Conceptual Medium: *Being/Affect*
Verb: *Connecting*
Visual Research Medium of the week: *Media of choice*
Exemplary Artist (a/r/tographers): *Dr. Mary Franco*
Readings: *Communities of A/R/Tographic Practice*; Irwin.
Thought in the Act: Preface: Manning & Massumi
Blogging as Art, Art as Research: Lucas Ihlein
Learning to be Affected: Matters of pedagogy in the artists' soup kitchen;
  Springgay & Zaliwska
*Notes on Interdisciplinary Methodology of Artistic Research: Visual
  Thinking Writing and Mapping*: Michelkevicius

Week 3:

Thematic Big Idea/Conceptual Medium: *A/r/tography*
Verb: *Transforming*
Visual Research Medium of the week: *Collage*
Exemplary Artist-A/R/Tographer: *Dr. Jo Stealey*
Readings: *Unflattening;* Pages 1–67; Sousannis
Collage as Analysis: Remixing in the Crisis of Doubt; Holbrook &
  Pourchier,
Thought in the Act: Thinking in Action (pg83–92) ((stop at proposition
  #3)) Manning & Massumi

Week 4:

Thematic Big Idea/Conceptual Medium: *Thing Power/Assemblage*
Verb: *Thinking With*
Visual Research Medium of the week: *Media of choice*
Exemplary Artist-A/R/Tographer: *Michael Murphy*
Readings: *Agential Realism, social constructionism, and our living rela-
  tions to our surroundings: Sensing similarities rather than seeing pat-
  terns*; Shotter
Thought In the Act Proposition # 3 & 4 (pgs. 92–98) Manning. &
  Massumi

Week 5:

Thematic Big Idea/Conceptual medium: *Play/Disruption*
Verbs: *Imagining*
Visual Research Medium of the week: *Shrinky Dinks*
Exemplary Artist-A/R/Tographer: *Dustin Yellen*
Readings: *Unflattening page; 67–113* Sousannis
Thought in the Act; Proposition # 5 & 8 Manning & Massumi

Week 6:

Thematic Big Idea/Conceptual Medium: *Relationship/Agency*
Verb: *Intra-acting*
Visual Research Medium of the week: *Media of choice*
Exemplary Artist-A/R/Tographer: The students
Readings: *Chapter 1 Plugging one text into another*; Jackson & Mazzei
Thinking with Agentic Assemblage in Posthuman Inquiry; Jackson &
   Mazzei

Week 7:

Thematic Big Idea/Conceptual Medium: *Connection/Multiplicity*
Verb: *Weaving*
Visual Research Medium of the week: *Fibers*
Exemplary Artist-A/R/Tographer: *Everyone!*
Readings: *Coming Alive in a world of Texture, pgs.1–30*, Manning &
   Massumi Text

Week 8:

Thematic Big Idea/Conceptual Medium: Emergence/Incorporeal
Verb: *Unfolding*
Visual Research Medium of the week: *Text as image*
Exemplary Artist-A/R/Tographer: *Everyone!*
(Watch) *Art Assignment; Combinatory Play*
Readings: *Toward an Aesthetic of Unfolding In/Sights through Curricu-
   lum*; Irwin
Unflattening; Sousannis, 113–152 Optional, read 157–169, Notes on
   theory and thinking about pages.

Week 9:

Thematic Big Idea/Conceptual Medium: *Assemblage pt 2*
Verb: *Emerging/Encounter*

Visual Research Medium of the week: *Oil pastels/tissue paper*
Exemplary Artist-A/R/Tographer: *Everyone!*
Verbal and Visual Readings: *No Title Yet, pgs. 59–80;* Manning &
    Massumi Text
Film as Research/Research as Film
Watch: https://vimeo.com/236295766
(Elizabeth Gilbert, you might have to copy and paste)
www.youtube.com/watch?v=86x-u-tz0MA

Week 10:

Thematic Big Idea/Conceptual Medium: *Reflection/Diffraction*
Verb: *Editing*
Visual Research Medium of the week: *iMovie*
Exemplary Artist-A/R/Tographer: *Dr. Amy Ruopp*
Verbal and Visual Readings: *Beyond an Easy Sense: A diffractive Analysis;*
    Mazzei
Thinking Without Method; Jackson
Amy's Screen Play
Watch: *Amy's Documentary; Portrait of an A/R/Tographer*

Week 11:

Thematic Big Idea/Conceptual Medium:
Verb: *Designing*
Visual Research Medium of the week: *iMovie*
Exemplary Artist-A/R/Tographer: You guys!
Readings: *Rendering Dimensions of a Liminal Currere*; Sameshima &
    Irwin
(copy and paste whole link)
www.edweek.org/ew/articles/2017/10/04/the-arts-havemuch-more-to-
    teach.html

Week 12:

Reading: Poetic Documentary and Visual Anthropology: Evoking the
    Subject
Weeks 13–15 were collaborative spaces of film and screenplay editing.

A Rhythm of Tension

Designed to introduce tension as a catalyst for a new thinking, this course
tangled with a number of theories and practices intended to intersect and
disrupt traditional notions of knowledge acquisition. Inspired by Deleuze

and Guattari, Mikulan (2017) offers an alternative lens to research, one that operates from an intuitive perspective and an exploration of the unknown, conceiving the research process as instaurative, "where the riddles and problems posed as theoretical questions demand not answers but the modulation of new problems and new questions" (p. 98). Like Mikulan, the intra-active inquiry which emerged in the course revealed new problems and questions catalyzing an ongoing deepening investigation into and under conceptions of what research is and does. The tension of the new, uncontrollable, unnameable, and productive discomfort became a threshold of difference and diffractive reading. "Insofar as an experiment involves a diffractive device, the experiment becomes a means of mutating concepts and resembling the world" (le Freitas, 2017, p. 742); in this case, the inner ontological worlds of student researchers. Le Freitas speaks of the notions of the experiment much in the way we construct theoretical play to practice through visualization and artistic mattering. It becomes a "cacophonous ecology" (Manning & Massumi, 2014, p. viii) where theory, ontology, and artistry explore the margins of knowledge acquisition and production. The student's ontogenetic explorations became an expression of their being, thus producing and expressing knowledge differently, unique to each of them. "Too often, writing stands to the side, outside the action, as though the 'real' work happened elsewhere, as though what writing was equipped to do with 'real' practices was merely to describe them" (Manning & Massumi, 2014, p. ix). The vignettes which follow illuminate the "real" work, a unique explorative encounter with theory. The voices of graduate student participants in the research course are invited into chapters through their own writings, makings, and musings, providing evidence of multiple and diverse responses and applications of theory as a conceptual medium. The blending of literacies illuminates the margins and in-between spaces of visual knowing rather than privileging traditional text-based knowing.

> The Arts produce and generate intensity, that which directly impacts the nervous system and intensifies sensation. Art is the art of affect more than representation, a system of dynamized and impacting forces rather than a system of unique images that function under the regime of signs. By arts, I am concerned here with all forms of creativity or production that generate intensity, sensation, or affect.
>
> (Grosz, 2008, p. 3)

My students embraced and participated with the expressive power of material, surrendering preconceived notions of how research is done. An important detail to include here is that throughout the course, I asked students to record their verbal/visual intra-actions with theory digitally, capturing their process and thinking in the moment. "Digital technologies let us do things differently-from digital humanities to 3D printing—and emerging analytical and visualization methodologies let us see differently. Doing and seeing differently means that we begin to interact with and experience the world differently" (Pendleton-Jullian & Brown, 2018, p. 29). So, it is with the world of research. These digital moments later served as data as they researched their own emerging realizations.

This chapter outlined one possible structure which set the stage for the thinking in the vignettes in Chapters 4–7. It affords a peek into such agential intra-actions, thinking, and doings and illuminates parallels to creative, visual, and artistic knowing. Continuous questioning, disruption of existing paradigms, and breaking down siloed learning/cross-disciplinary/knowledge reciprocity challenge the notions of privileging verbal text above all others.

References

Barad, K. (2009). https://quod.lib.umich.edu/o/ohp/11515701.0001.001/1:4.3/-new-materialism-interviews-cartographies?rgn=div2;view=fulltext

Grosz, E. (2008). *Chaos, territory, art: Delueze and the framing of the earth.* Columbia University Press.

Jackson, A. Y., & Mazzei, L. A. (2013). Plugging one text into another: Thinking with theory in qualitative research. *Qualitative inquiry*, 19(4), 261–271.

Le Freitas, E. (2017). Karen Barad's quantum ontology and posthuman ethics: Rethinking the concept of relationality. *Qualitative Inquiry*, 23(9), 741–748.

Manning, E., & Massumi, B. (2014). *Thought in the act: Passages in the ecology of experience.* University of Minnesota Press.

Mikulan, P. (2017). *Pedagogy without bodies* (Doctoral dissertation, Education, Faculty of Education).

Pendleton-Jullian, A. M., & Brown, J. S. (2018). *Design unbound: Designing for emergence in a white water world* (Vol. 2). Infrastructures.

Sousanis, N. (2015). *Unflattening.* Harvard University Press.

Springgay, S., & Zaliwska, Z. (2015). Diagrams and cuts a materialist approach to research-creation. *Cultural Studies↔ Critical Methodologies.* https://doi.org/10.1177/1532708614562881

St. Pierre, E. A. (2017). Deleuze and Guattari's language for new empirical inquiry. *Educational Philosophy and Theory*, 49(11), 1080–1089.

Suwa, M. (2001). Constructive perception in design. *Computational and cognitive models of creative design*, 227–239.

4 A Matter(ing) of Ethico-onto-epistem-ology

As noted in Chapter 3, the course was designed as a rhythm; offering readings around poststructural, posthuman, and (feminist) new materialism theories. These were engaged within and through visualization and creative practices. An integral part of the conceptualization for these encounters was to have students examine and explore their own philosophical roots. An emphasis in research often rests on notions of epistemology, with a curious disregard for ontology (St. Pierre, 2018). While an entire book could be written about this subject alone, I leave it to the reader to investigate and recognize that this is an incredibly complex topic that I have not taken lightly. As such, I felt it was critical for students to examine their own philosophical dispositions. Afterall, these ways of knowing impact how we navigate the world and interpret what we perceive. Understanding one's ontological, epistemological, and axiological dispositions sets a stage for an informed and "response-able" (Haraway, 2016) foundation for conceptual theorizing with(in) and about research practice.

This first vignette focuses on a first-semester Art Education doctoral student and follows her transformation as she explores her ethico-onto-epistemological[1] dispositions. The first semester of any doctoral program can be a little bit shocking. There is imposter phenomenon, culture shock, and often an overwhelming sense of *what am I doing here*? Graduate work is a place of tension. It should be! If it were easy, what would we learn? In her first semester, Lauren enrolled in my Artistic Research course and simultaneously, an introductory qualitative research course which outlined the more traditional tenets of qualitative methodologies. Already, the potential for tension was present as my course was paradigmatically opposite to the content being presented in the other course (all valuable here, I do not mean to dismiss other methods and ways of doing research). Coming from an elementary art education background with an additional interest in educational leadership and administration, Lauren entered the class as a self-identified systems thinker. She conceptualized herself as a teacher and administrator,

DOI: 10.4324/9780367823818-4

but not as an artist. Her sense of creating had been hibernating. Like many art educators in the field, she poured her efforts into students, curriculum design, and things of school, forgoing the visual side for "another time". The first few weeks of the class were a challenge for her. The nature of the class being to deliberately introduce tension and disrupt habits created tremendous discomfort for her. As we pored over initial readings, the anxiety present in her writing and inter/intra-actions[2] with theoretical concepts was palpable as she explored concepts visually. Yet, she persisted.

> *My tension piece I chose came from this reading. On page 25 Sousanis writes, "**Disrupting these deeply ingrained patterns takes a profound nudge . . . a rupture in experience, illuminating boundaries and the means to transcend them**" (Sousanis, 2015, p. 25). I sat with that thought for a moment. It was hard and rather emotional, because the process one must go through to do that, to disrupt deeply ingrained patterns is intense and hard and painful and emotional and should cause heaviness, tension, and sadness.*
>
> *My motivation for the way I visually reacted was to put those words front and center, because when you begin to see ideas and people from new perspectives even if it's not your own, you can't just forget it. It works on you and breaks down walls and barriers and belief systems and it ruptures your core. At times, it hurts. I ripped the paper because it felt like what it feels like to go through the process. You get to the core of who you thought you were and you have all these pieces you have to put back together. In my reaction I chose staples. Staples are like having legit surgery. The surgeon doesn't bring out the staples unless it is a large wound that won't stay together any other way. The staples are like the battle wound. It's my way of saying I persisted. I had guts to go through it, and have the scars to prove it. In the end if you do the work, something beautiful and transformational emerges and that is why I drew the butterfly. The butterfly takes on a miraculous transformation all its own and is the perfect metaphor to end on.*
>
> *(BLOG POST SEPTEMBER 2017)*

The readings paired with visual explorations were pushing up against her own artistic insecurities and sense of self. Her initial visual response to readings very much reflected her systems thinking.

Suddenly, she was (as were others) asking questions like, "Do I want to let go?" and, "What do I know? Is what I know real?" We build our worlds around what we believe we know to be true, often with very little internal interrogation of what this is and where it comes from. However, when one pushes up against long-standing belief systems, challenging their function, legitimacy, truth or what have you, the situation gets tense! This disruption requires courage and perseverance. The potential of this disruption to generate cracks in long-held beliefs opens up new spaces for discovering, knowing, and understanding at the core. Thus, to dive deep into all the theoretical explorations we were going to work with, I felt it was extremely important for all students to understand the significance of the philosophical roles of ontology,[3] epistemology,[4] as well as axiology[5] within their own research practice. These three philosophical lenses, in part, shape and influence how we engage with inquiry.

Since an essential practice of the course was to make inter and intra-actions visible, the very first significant visual assignment students received was to explore and express their ontology in visual form. This would become the first layer of a three-part assignment exploring ontology, epistemology, and axiology, a conceptual self-portrait. We spent time in class discussing beliefs about being in the world. What is true? What do each of us believe and why? Where did these foundational beliefs come from? Recognizing the potential paralysis some students might experience when being asked to create a philosophical self-portrait, I broke down each visualizing layer into smaller chunks. As we began with ontology, we generated (without judgment) "I believe" statements. Our beliefs are the lenses through which we see when we come to know something. These statements served as idea seeds (Ruopp, 2019) or points of departure for visualization. We analyzed these statements noting the underlying assumptions and actions. In the writing of these "I believe statements", I asked, "what actions might emerge from this and how might these actions matter as a visualization?" As researchers of any discipline, it is critical to have a sense of this foundation as it interweaves into everything we do. These belief statements inspired metaphors and visualizations, putting the verbal and visual texts to work through each other, in service to one another.

Layer 1: Ontology

> *I have looked up the word ontology no less than 50 times in the course of the past few weeks. It is an abstract concept that has been a little tricky for me to grasp. The idea of what I believe about being? I'm not sure that I had ever really given it that much thought. I liken it to what my core values are.*
> (Blog Post Early September)

In thinking so deeply and philosophically about their conceptualizations of being in the world, I asked the students to call upon their artist selves. Through the artist lens, they had to give themselves permission to play.

Figure 4.1 Early intra-actions

Figure 4.2 Intra-action with Nick Sousanis (2015, p. 25) UNFLATTENING by
    Nick Sousanis, Cambridge, Mass. Harvard University Press, Copyright
    2015, by Nick Sousanis. Used by permission. All rights reserved.

To be curious, to make mistakes, and to be messy. For some, yet again, this caused extreme discomfort. This required them to shift from a product focus to a process emphasis and stay present in the moment to allow for the emergence of imagery that might inform them about their own ontological dispositions.

> *Speaking of disruption, the portrait situation has put me through the ringer this week . . . . It's been a long time since I have felt that amount of discomfort surrounding projects . . . Cheers to that, maybe????!*

(Blog Post Early September)

I asked them to recall the pallet of verbs from the exercise we explored during the mapping of the actions and doings of A/R/Tography, reminding them that there are actions that emerge from and influence our beliefs and behaviors about how we are in the world. How might these actions matter as a visualization? What verbs are present in your belief systems? Some of the actions that emerged in Lauren's "I believe statements" were: voicing, connecting, learning, growing, searching, and interweaving. Lauren contemplated and thought out loud, the metaphor of weaving arose. She makes mention of time and interconnection. She began to recognize that ontology is cumulative and woven through time and experience.

> *I printed this*[her belief statements] *out in multiples and aged the paper by crunching it up and re flattening it. After doing this a time or two, I used some ink in a spritzer bottle to stain the paper in varying hues. I made this decision because your ontology isn't new and shiny. It's worn and aged and weathered from experience.*

(Blog Post, Early September)

> *On page 158 when he (Sousanis, 2015) uses the phrase, "**a glimmer of possibility often obscured**" and again on page 166, when he references "**dimensions we can't experience are curled up tightly within those we can**". The idea then for me is that this process of an aesthetic awareness of the in-betweenness is exactly what our portrait of being in the world is.*

(Blog Post Mid September)

Through the act of doing, remaining present, and slowing down, Lauren was able to notice intersections and connections with/in/between the actions of her beliefs. This facilitated a productive dialog between the material and her own conceptualizing. In thinking with both verbal and visual literacies,

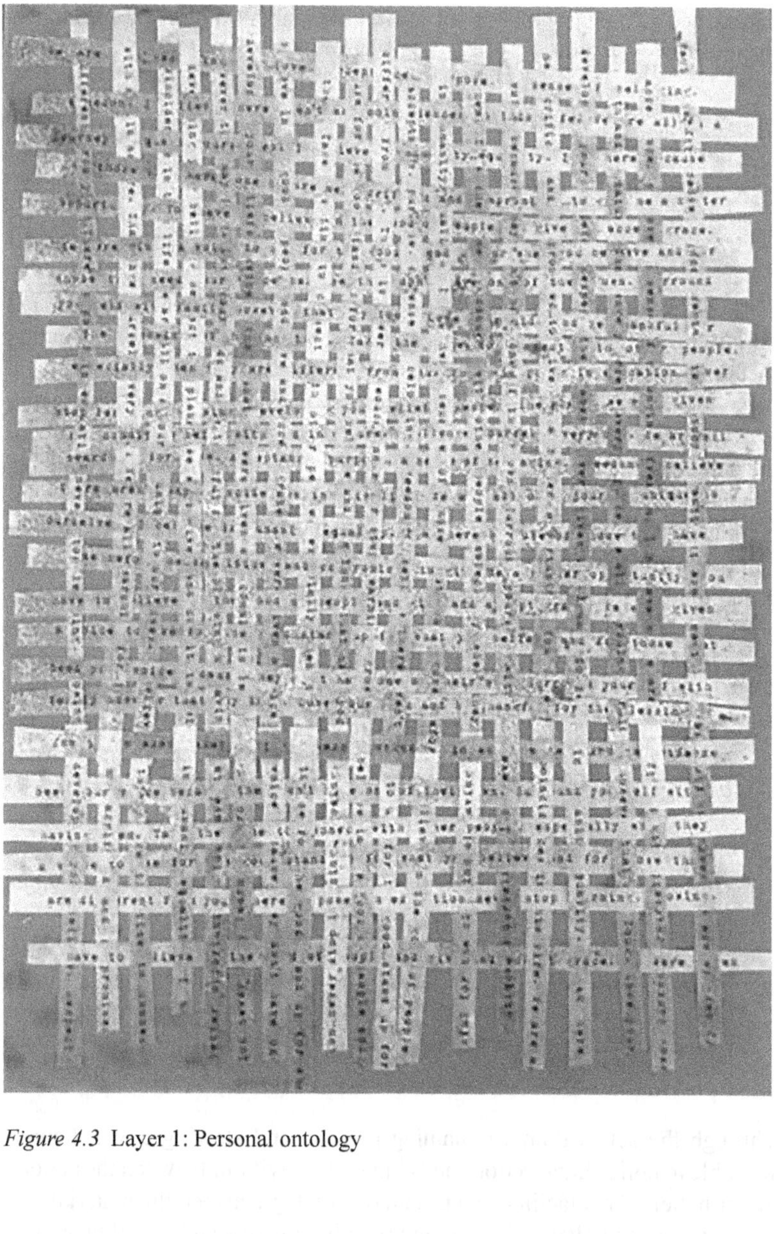

Figure 4.3 Layer 1: Personal ontology

Lauren was able to successfully articulate and produce a representation of her ontology.

Sharing

An important aspect of this process was to come together and share realizations as each individual's ontology unfolded. In doing so, the diversity of process emerged which fueled discussion around theoretical connections. To their surprise, this often strayed from the theories and conceptual frameworks they had previously claimed and in which they had become comfortable. How might these realizations impact their research questions or modes of inquiry? An eager sense of anticipation for the development of the next layer was present. What began as apprehension and tension was now being beautifully articulated in the collaborative setting. It is important to note that I also participated in all of the assignments I gave them as a way of participating as a co-learner in an A/R/Tographical environment. In doing so, I likewise pushed up against my own habits and persisted through tensions.

Layer 2: Epistemology

The next layer dealt with epistemology. How do you know what you know? Cotter (2017), quotes Susan Sontag when discussing art as "mainly a form of thinking" and "each work of art gives us a form or a paradigm or model of knowing something, an epistemology", she further discusses

Figure 4.4 Students exploring and discussing ontological portraits

Figure 4.5 Students exploring and discussing ontological portraits

the importance of material inquiry and "art as a site of thinking" (p. 11). After the initial success and confidence boost of the ontology layer, students became attached to the product of their hard work and thinking. Transforming this to include another philosophical expression reintroduced tension and anxieties. Concern over "messing up" the first layer was expressed. I asked them to sit with this and wonder. I reminded them to reorient and focus on the process and that this investigation was ongoing and fluid. At this point of the semester, we had also read several additional theories (rhizome, diffraction, and assemblage to name a few) and intra-acted with the actions of those theories. In essence, they had additional conceptual tools on their pallette to explore. For Lauren, this was initially a complicating factor:

The beginning is also very difficult for me. I have to work it all out in my head before I can start in on the work. I felt really uneasy about covering my weaving. It was pretty laborious and I thought it explained my ontology really well. I decided that for this layer I wanted color because an explosion of color aligns with how I know things. The origin of my knowledge has hit in these bursts throughout my life. If I think about it in regards to graduate school though, I can almost pinpoint each stroke of color as an acquisition of something new. Once I saw the paint over the textures of the weaving, I knew I had made the right choice. The weaving took on a new purpose being the base layer and added an element of depth and texture that allowed the paint layer to have some interest; much like our ontology continues to be influenced by our epistemology and our epistemology is organized through the perspective of our ontology. The two layers in my piece inform each other. All of the circles are representative of the cycle of knowledge. They float around in the painting as if they could change or float away at any moment.

(Blog Post, Mid September)

As she contemplated her process in practice, the intersections between the two philosophical dispositions surfaced. During discussions in class, I introduced Barad's notions on ethico-onto-epistemology more in depth as the entanglement of these philosophies "mattered" before our eyes.

It becomes less about what you expect to find as far as the truth, and more about how you interact with all of the pieces through and with, in finding what the experiences want you to find. The passage I chose speaks to this idea of using diffractive analysis counter to that of traditional coding causing the researcher **"to thread through or 'plug in' data into theory into data resulting in multiplicity, ambiguity, and incoherent subjectivity . . . it is . . . a moment of plugging in, of reading-the-data-while-thinking-the-theory, of entering the assemblage, of making new connectives."** *With each reading, I understand a little more, build a little more on my own thinking and can actually . . . dare I say . . . see where my own onto-epistemological self is starting to camp out!"*

(Early October, 2017)

Layer 3: Axiology

Exploring axiology was the final layer. We began our discussion contemplating various definitions of axiology including the following:

To value is to set priorities. It is to choose one thing over another. It is to think about things in relation to each other and decide that one is better than the other. It is to decide what is "good". All persons assign higher value to some things and lower value to others. People assign these valuations in a consistent pattern that is unique to them. This valuation process is actually one's habit of thinking. It involves filtering, processing, storing, and analyzing data. It includes thinking about objects, discerning the different aspects of things, making judgments, and choosing.

(www.cleardirection.com/docs/axiology.asp)

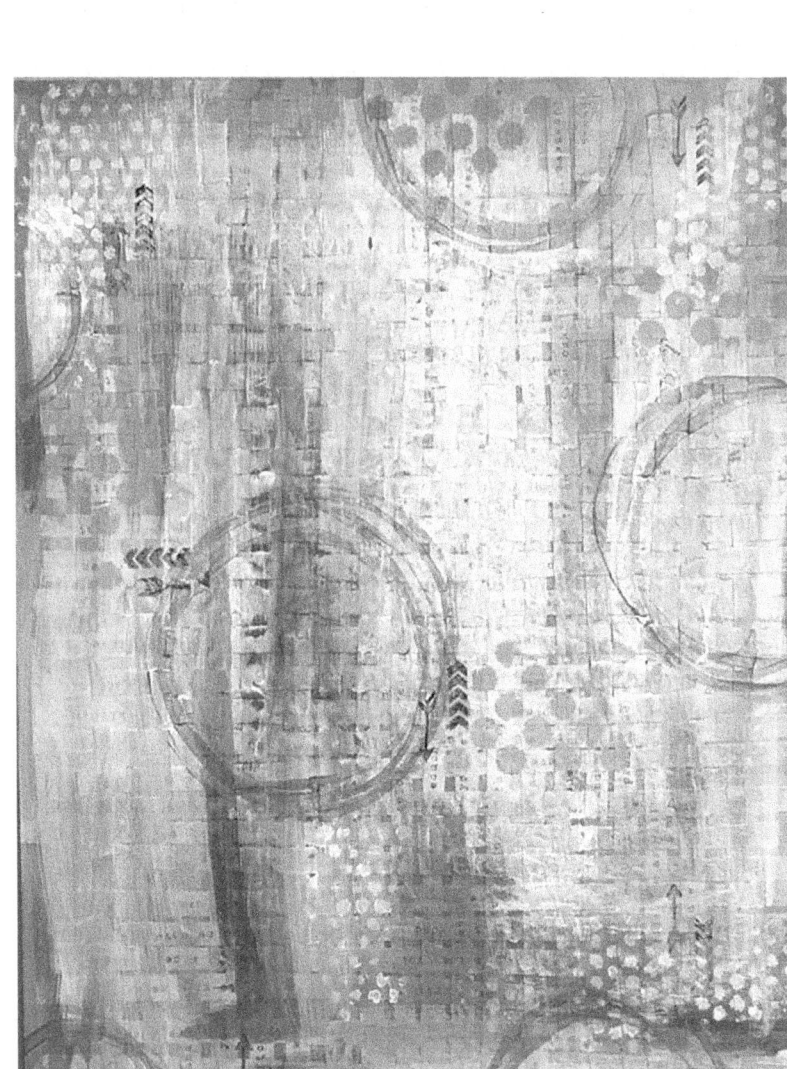

Figure 4.6 Layer 2: Epistemology

Noting the researcher as the central focus of most of the definitions we found, we looked to and reviewed Barad's ethico-onto-epistemology and Haraway's notions or response-ability.

As we rethink matter, we must rethink the empirical (about knowledge) and ontology (about being), and the classical division between the two begins to break down, hence, Barad's (2007) new concept onto-epistemology and another, even more indicative of this new work, ethico-onto-epistemology, which makes it clear that how we conceive the relation of knowledge and being is a profoundly ethical issue, as is the relation between the human and the nonhuman.

(St. Pierre et al., 2016, p. 99)

That week we explored theories around "thing power" in the class as a catalyst for thinking through posthuman notions of object agency. The exercise specifically challenged students to surrender to the material to learn from it. We contemplated how this applies to ethical research and what decentering the human might produce in relation to research methodology and knowledge production. Kassahun (2020), reflecting on St. Pierre's work offers,

One characteristic of post-qualitative research orientations is the decentering of the human researcher by consciously avoiding, for a much as is possible, predefined method but rather to begin with an intelligible encounter that compels and presses the researcher to think with the subjects of inquiry.

(St.Pierre, 2018) (p. 395)

As Lauren approached her axiology layer, she deeply contemplated several quotes from the reading.

I was drawn to a specific interpretation of the theory by Max Scheler, in which he determined that, "values are instantiated in physical objects but they are not themselves physical objects." (www.newworldency clopedia.org/entry/Axiology) I found this to be especially important in understanding that in order to explain my values, I objectify them, but they are not the objects themselves. Once I understood this idea of what has value to me, it was easier for me to understand how I would complete my final layer. I set off to discover the sort of aboutness of my value set. I immediately became fearful of putting anything else onto my portrait. It didn't really have anything to do with messing it up or having it not look right, but more to do with the time and intentionality I had with the other two layers. Even though with layer 2, I completely

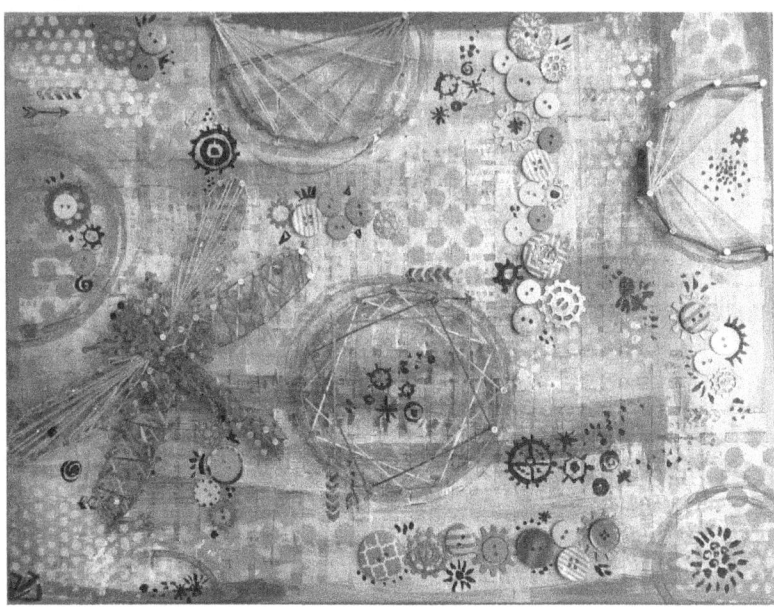

Figure 4.7 Layer 3: Axiology

> *transformed layer 1, the integrity of layer 1 still exists and is a large part of the piece. I felt really strongly about the concept that all the layers should exist together because each one has particular "value."*
>
> (Blog Post, Mid October, 2017)

> *As I even type, I'm reminded that I am the one choosing the words and symbols which alone are not what I value but the vehicles for which I give value agency.*
>
> (Blog Post, Mid October, 2017)

Contemplations

Our philosophical lenses "matter" quite literally. They impact the assumptions and choices researchers are making when designing and conducting research and inquiry. Data, analysis, interpretation, and even what to study is interwoven through these lenses. Interrogating and disrupting long-held beliefs, while potentially uncomfortable, makes space for something new to emerge. It empowers researchers to consider something different than before, a sense of response-ability for and with all human and nonhuman entities. "Response-ability" refers to one's ethical sensitivity and the ability to respond accordingly. Haraway defines "response-ability" as "cultivating collective knowing and doing" (Haraway, 2016, p. 34). This was unfolding throughout this process and evident in our class discussions.

Approaching this exercise from a space of emergent possibility offered students from all disciplines and backgrounds multiple entry points into this process of discovery. Brown (2017) quotes Nick Obolensky to describe Emergent as "the way complex systems and patterns arise out of a multiplicity of relatively simple interactions" (p. 13). There is little that is simple about understanding one's ontology, epistemology, and axiology. As single concepts, they are already quite complex. Once combined and realized as Barad's ethico-onto-epistemology, the entanglement and inseparability are even more challenging to articulate. While we were only just beginning to scratch the surface of these concepts, they remained present in the students' thinking throughout the class and beyond, as they continued down their now altered research paths.

The process of reading, drawing, speaking, and writing polyphonically produced dynamic unfoldings of awareness, conscious realization, and shifts in practice. A final quote from Lauren:

> *The passage I chose was* on page 64 of Manning & Massumi; **". . . painting is not about seeing. It is felt, it touches, it moves, it resonates. To paint the outside is about feeling-with, a thinking-feeling that occurs in a relational field, across works in the making"** (Manning & Massumi, 2014, p. 64). *On the following page they even clarify this by extending the thought;* **"Thought gathers in the work . . . not into language, but in painting"** (Manning & Massumi, 2014, p. 65). *The idea of think-feeling and thought-felt were ideas that were new and insightful. I could rally around that idea of seeing the painting not through the content, but instead, the content emerging because of what the artist was feeling, and without thinking.*

Notes

1 The notion of "ethico-onto-epistem-ology" was first coined by physicist-philosopher Karen Barad to point at the inseparability of ethics, ontology, and epistemology when engaging in (scientific) knowledge production, with scientific practices, and with the world itself and its inhabitants—human and non-human beings that intra-actively coconstitute the world (Barad, 2007, p. 90). https://newmaterialism.eu/almanac/e/ethico-onto-epistem-ology.html

2 Intra-action is a Baradian term used to replace "interaction", which necessitates pre-established bodies that then participate in action with each other. Intra-action

understands agency as not an inherent property of an individual or human to be exercised but as a dynamism of forces (Barad, 2007, p. 141) in which all designated "things" are constantly exchanging and diffracting, influencing, and working inseparably. Intra-action also acknowledges the impossibility of an absolute separation or classically understood objectivity, in which an apparatus (a technology or medium used to measure a property) or a person using an apparatus are not considered to be a part of the process that allows for specifically located "outcomes" or measurement.

SYNONYM: agency, animacy, connectivity
ANTONYM: disconnection, inanimacy, objectivity, singularity, vitalism
HYPERNYM: phenomenology, connectivity, acknowledgement, respect
HYPONYM: interaction, exchange, movement
REFERENCES
Barad, K. (2007). *Meeting the universe halfway: Quantum physics and the entanglement of matter and meaning*. Duke University Press.
https://newmaterialism.eu/almanac/i/intra-action.html

3 Ontology is the "study of being", which is concerned with what actually exists in the world about which humans can acquire knowledge. Ontology helps researchers recognize how certain they can be about the nature and existence of objects they are researching. For instance, what "truth claims" can a researcher make about reality? Who decides the legitimacy of what is "real"? How do researchers deal with different and conflicting ideas of reality?

To illustrate, realist ontology relates to the existence of one single reality which can be studied, understood, and experienced as a "truth"; a real world exists independent of human experience. Meanwhile, relativist ontology is based on the philosophy that reality is constructed within the human mind, such that no one "true" reality exists. Instead, reality is "relative" according to how individuals experience it at any given time and place. https://i2insights.org/2017/05/02/philosophy-for-interdisciplinarity/

4 Epistemology is the "study of knowledge". Epistemology is concerned with all aspects of the validity, scope, and methods of acquiring knowledge such as (a) what constitutes a knowledge claim; (b) how can knowledge be acquired or produced; and (c) how the extent of its transferability can be assessed. Epistemology is important because it influences how researchers frame their research in their attempts to discover knowledge. https://i2insights.org/2017/05/02/philosophy-for-interdisciplinarity/.

5 Axiology is the study of value or, more adequately, theory on the nature of value. In plain-English, what is good (or bad) in life and what do we find worthy.
Axiology incorporates ethics (theory of morality) and esthetics (theory of taste and of beauty), as well as other forms of value. Asking what "ought to be" is axiological. https://i2insights.org/2018/05/22/axiology-and-interdisciplinarity/

References

Barad, K. (2007). *Meeting the universe halfway: Quantum physics and the entanglement of matter and meaning*. Duke University Press.
Brown, A. M. (2017). *Emergent strategy*. AK Press.
Cotter, L. (2017). Reclaiming Artistic Research–First Thoughts.... *MaHKUscript. Journal of Fine Art Research*, 2(1).

Haraway, D. J. (2016). *Staying with the trouble: Making kin in the Chthulucene.* Duke University Press.

Kassahun, W. (2020). "Becoming-with bees": Generating affect and response-abilities with the dying bees in early childhood education. *Discourse: Studies in the Cultural Politics of Education, 41*(3), 391–406. https://doi.org/10.1080/0159 6306.2019.1607402

Manning, E., & Massumi, B. (2014). Preface. In *Thought in the Act* (pp. vii–ix). University of Minnesota Press.

Ruopp, A. (2019). Portrait of an A/R/Tographer: Theory as Conceptual Medium. *International Journal of Education & the Arts*, 20(9).

Sousanis, N. (2015). *Unflattening.* Harvard University Press.

St Pierre, E. A. (2018). Post qualitative inquiry in an ontology of immanence. *Qualitative Inquiry.* https://doi.org/10.1177/1077800418772634

5 Marching with The Emerging Artist/ Researcher/Teacher

As in Chapter 4, *the voice of the student is shared in italics,* **citations are in bold italics,** and my own voice is standard text. Varying the visual aspects of the texts, it is my hope to afford individuality to each voice.

> *I am seeking ways to teach the unspeakable, unspoken, and unthought of local history, particularly as it relates to the black experience in social studies education.*
>
> (TWK, personal communication, 2019)

The composing of this chapter for me, as a cis gender white woman, writing about the experiences of a cis gender black woman, created notable tension for me. I am acutely aware of my privilege and want to be certain that my voice is in service of her work. While my attention centers on the emergence of her visual voice and how it empowers her thinking and research, I am deeply aware that her topics for research are poignantly interconnected to issues of systemic racism. Her visualizations of theory and their connection to her research and current events were and are profound.

ME: Good morning. I am reading your blogs again and over and over, you speak to your stick people "marching". They move differently in different entries, and they visually transform overtime, from early almost abstract wisps into powerful groups with intention, wanting to be seen. I am reading this as I see over and over again on social media and on TV etc. people marching. More and more people marching. It's almost as if your stick people have mattered into the world.

TWK: *They do a lot of emotional work for me. Some days I am starved to work for and with them. Today is one of those days.*

This book is being written during COVID-19. Concurrently, protests and marches are happening all over the country. As I re-read through the blogs

DOI: 10.4324/9780367823818-5

Figure 5.1 Marching stick people

of this particular student, the parallels and intersections between the concepts she was working through and present-day materialization of them affords me yet another lens to contemplate her powerful thinking and realization through visualization. I learn from her and her visualizations each time I encounter them.

A Little Bit of Context

I met TWK when I was a guest instructor in a postqualitative research course prior to her participation in the Artistic Research course. At that time, I was ABD, and she was completing her master's degree and was applying for the PhD program in Social Studies. I had been invited to share my experiences as a visual researcher and A/R/Tographer. As I went through my presentation with the class, I noted some of the students were much more interested than others. Challenging conventional qualitative research requires finesse. The social sciences have developed a considerable repertoire of research methodologies which offer concrete structures and paths to follow. I was not there to tear those down or suggest that they are substandard. They have their place. With that, the security of method has deep roots sometimes limiting conceptualizations of research. How does one discover the unknown if you already know what you are looking for? Disrupting the familiarity of the tried-and-true methods with a clear path for ambiguity and tension was met with some skepticism. Therefore, I was diligent in offering multiple artistic research examples and visual thinking strategies for them to directly encounter and experience for themselves. The lived a/r/tographical engagement and community discussion afforded insight and sparked interest in many of them. At the end of the class, one student remained behind. It was TWK. I will never forget the look on her face. She was sitting with a quiet but intense appearance. She exclaimed, "You think like I do!" The excitement was palpable for me as well as I recognized the light in her eyes that evening. Recognition, maybe hope, and validation. After 18 months, she enrolled in my Artistic Research course. TWK began the course feeling shy about her artistic skills and with significant apprehension about her ability to express ideas visually. This chapter illuminates the unfolding and celebration of her visual voice, hence the significant presence of imagery. Her persistence through tensions and willingness to embrace newness empowered her thinking in new ways, revealing the power of image to provoke thinking and serve as a vehicle for research findings.

The artifacts shared in this chapter are mined from her blog posts as well as some personal communications. This includes her thinking during her time in the Artistic Research course and samplings of her continued work spanning the following year. Her blog writings are accompanied by drawings and material explorations from the readings in class. Her blogs are also rich with citations in conversation with her verbal and visual/material musings. They are beautifully fluent conversations. As such, this chapter is constructed to read much the way her blog does. I add to the conversation my own reflections, musings, and insights. This chapter invites extensive dialogue with her visualizations. As you, the reader, engage with this, I invite you to add your own visual and verbal musings to the text. On with the story!

Week Four: Confronting Tensions

Four weeks into the semester, I had just finished sharing with the students that they would be creating a short documentary film from their raw video data. As outlined in Chapter 3, students were video recording their visualizations and intra-actions with theory. The goal of revisiting, researching, and editing the raw moments of their process would be to illuminate their newly emergent realizations on how visual literacy empowered their thinking and put theory into literal practice as research. Although TWK worked visually as she researched, mapping and drawing out histories and lineages, she did not yet identify as an artist.

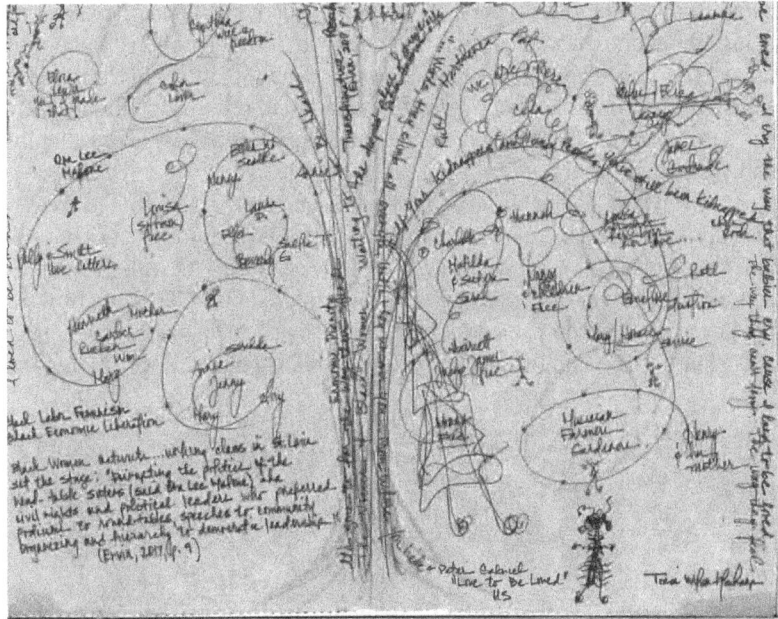

Figure 5.2 Mapping thinking

TWK silently turned to me with a look of terror crossing her face. The tension of drawing was already challenging her notions of self and the validity of visual understanding. She, like so many, equated artistry with technical ability. She was being asked to work in a completely new territory and the discomfort was real. I smiled, recognizing this feeling and confirmed it was normal to feel uncertain. I assured her she would be fine. Initially, she was not convinced.

Productive Confusion

St. Pierre (2016) discusses the need for **"productive confusion of not knowing what to do"** (p. 7). This confusion, which often dialogs with tension invites attention elsewhere if we are willing to engage with it and surrender to it. Each encounter with artistic materials was designed to invite visual thinking and doing into serious but experimentally playful conversation with theory. "Experiments are risky creative events that reassemble [their inner] world" (Le Freitas, 2017, p. 742). Being with(in) a state of tension over time reveals how we navigate encounters with knowing and being from an ontological perspective. "We seldom talk about the nature of being, about ontology, because it is assumed in this structure that to be is to know" (St. Pierre, 2017, p. 1080). However, I offer that to *be* (with)in tension, a diffraction of how knowing might unfold to reveal something different. This opens a new space where theory and practice *matter* in meaningful and significant ways

As noted in Chapter 4, an important assignment in the course was to visually explore their ontology, epistemology, and axiology as a cumulative single piece over the first 6 weeks of the semester. As with their intra-actions with theory, this process was also captured on video as there were readings which accompanied their visual musings. As the work developed, a/tension to the entanglements of the three philosophical dispositions was

Figure 5.3 Exploring onto-ethico-epistemology

explored through Barad's (2007) notions of onto-ethico-epistemology. In noting the distinct differences, but also the absolute intersections, it transformed the conceptualizations of theoretical engagement and its relationship to research. The three lenses of the artist, researcher, and teacher are discrete yet interconnected, just as the ethics of being and knowing are. By illuminating her philosophical dispositions, she was able to visually work with theory in a deeply visual and productive manner that allowed her to articulate new realizations. This fertilized the seed of her artist lens as she surrendered and gave permission for imagery to emerge. Her artist identity was beginning to root itself in her awareness;

Mess as a Formative Force of Thought

Actually "doing" my ontology keeps unraveling the depth of my epistemology, revealing their deep companionship.

*I suddenly had a moment of affect—a shift in that brief in-between rising and getting up: "Oh . . . I get my ontology! It clicked. The biggest tension I experienced was the messiness of the process and not knowing what I was going to do with all the scraps of paper I had dug out in a heap. I actually ironed a stack of crinkled and balled up scraps of wrapping and tissue paper—hoping to be thunder-struck with ideas. I identified with the passage noting **"We see ourselves not as distinct and divided or fixed and grounded . . . we draw upon the Deleuzian concept of the fold to conceptualize how our multiple selves are in constant relation and mutation"** **(Holbrook & Pourchier, 2014, p. 755).** The little girl part of me, I think, was awoken in this process and she went to work! She is a very different self than the teacher, researcher, and artist-in-the-making. My frustration was not being able to create in terms of drawing more life-like images because I don't have that kind of skill-set. To compensate, I take lots of photos of my process and musings as they come together. I felt the frustration of "**I can't figure out how this is going to come together, but I feel like it will come as I work" (Holbrook & Pourchier, 2014, p. 756).** And I believe it all did come together and I was surprised at my engaged interest, curiosity and willingness to bend, play, and lay scraps over and over again seeking to let the **"images to the work of fragments of sentences, fragments of thoughts, of memory."***

(Holbrook & Pourchier, 2014, p. 756)

I sense my ontology is shaped by homemaking or creating comfort? Or maybe it is my attention to creating a space that is aesthetically pleasing to me. Maybe it is the little girl in me who still relishes playing house and making doll clothes out of paper, kleenex, and scraps?

The pieces of my portrait are coming together in pieces. One part ontological via the Red Fabricated Lady; the second visits via my epistemological threshold as the Pink Lady—she enters and exits through multiple openings through via threshold which acts as "a passage way . . . attached to other things different from itself" (Jackson & Mazzei, 2013, p. 264). The third layer is home to my axiology which acts as a volcano of melted ore mixing strategically with the flexibility of hot glass twisting, wading, and moving "thing-power"—a.k.a. my stick people who resist, march, and demand through mouth and voice.

In investigating a/r/tographically and intra-actively, TWK interweaves her creative visual play with intersections of the reading and personal meaning. Her reviving the feeling of play from childhood likewise invited in an additional element of perception and interaction, giving her permission to be messy. "Playing encourages self-discovery, an insight into self, surroundings and relationships" (Szekely, 2015, p. 13). Moving from her exploration of her ethico-onto-epistemology, her engagement with image and material becomes increasingly central to her investigations. Now concepts of assemblage emerge as she overlays each image in her process with a relevant citation from her readings. By week nine, she is exploring notions of affect and assemblage, her voice almost more visual than verbal while in conversation with specifically selected citations.

She opens her thinking with:

"If white is painting-thinking coming into its own, the coloring of the process is "affect" (Delueze, 1986, p. 118) . . . . Color is this thinking's affective accompaniment. Its feeling friend. Painting: thinking-feeling. Not imagining" (Manning & Massumi, 2014, p. 80).

Honestly? I had a dream of being a machine and plugging in various scenes of juxtaposition and then sitting back and observing, taking notes, talking to myself and then on to the next. The dream went on and on . . . and then a flash of the words, "brushstrokes across the paper" rotated on the "screen" in my dream like the words that scroll across the bottom on CNN.

I woke up desiring to paint but what.. I had no idea? I tinkered and played for what seemed like forever. I cut up white squares of trace paper and doodled with colors wondering if I could create a visual "act [of] continuously disappearing" (Manning & Massumi, 2014, p. 72). Could I layer my stick people with colors and somehow use additional sheets of transparency paper to create a disappearing fog? Nah. Nope! Didn't like that. Sucked. Then, I cut up lots of pieces of white paper. I tried to fold the paper and clip out a person that continued [to]unfolded together. Somehow I missed the memo on how to do that? After a while (probably because I am persistent) the piece of paper emerged into something. What I didn't know. I just figured they'd be useful at some point. I thought of Dustin Yellin and his childhood habit of curating collectibles left by nature or human absence . . . I figured that these white cut out images might lead me to create something?

In the margins of Irwin's (2003) Toward an Aesthetic of Unfolding In/Sight Through Curriculum" (p. 64) I scrawled:

"In my dream I was plugging in scenarios in my juxtaposition machine, observing each output like I was adoring a painting or a drawing. There were folds within folded, openings and spaces of wonder and beauty and a multiplicity of views.

There are spaces of white that interchange in and thru the machine prompting viewing options. There was a rhythm to it. My challenge was to keep plugging in the machine (the juxtaposition vase of roots in my epistemological layer). . . . "

I played with fabric and the gluey-paper . . . Then I took up my paints and began to paint . . . aimlessly and cluelessly. I was meandering, it was late, but I wasn't ready to give up. My senses seemed still willing (tired as they were) willing to probe a little further. Irwin (2003) says this "aesthetics awareness is open to wonder while suspending belief and trusting uncertainty" (p. 64).

As usual I started again making my squiggles of stick people. But with paint, I have less control over the width of the lines. Lots of them are in blue, then a green, then more blue, smatterings of pink.. and I thought. Wow. I'm going nowhere with these colors. What about white? What can I do to make my parade of constantly marching stick people appear to disappear . . . to be in the background? How do I begin a process of "underpainting" (Irwin, 2003, p. 66).

Assembling Moments of Inspiration

What follows is the manifestation of her thinking above. The selected citations are poetic in their assembling in conversation with the imagery as it emerges.

Face becoming imperceptible; no name. No title-yet.
(Manning & Massumi, 2014, p. 76)

disrupting and surprise . . . allows aesthetic knowing to emerge.
(Irwin, 2003, p. 67)

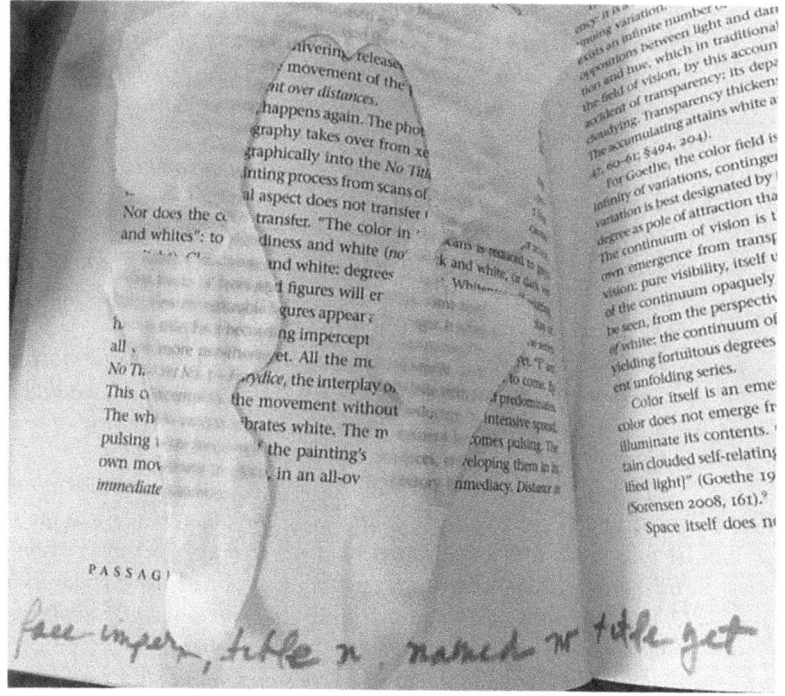

Figure 5.4

surrendering to the unknown often brings disruption and surprise.
(Irwin, 2003, p. 67)

Thinkfeel the indiscernible in action, the coming imperceptible of the I,
the trembling of an outside that is never quite there for the seeing. That
is not to say that the figure is actually excluded. The figure is in there,
in the act of continually disappearing.
(Manning & Massumi, 2014, p. 72)

Aesthetic awards open to wonder and expression.
(Irwin, 2003, p. 67)

wonder and surrender [allow] being attuned to what is unfolding.
(Irwin, 2003, p. 67)

Figure 5.5

and aesthetic of surrender is complemented with an affirmative aesthetic.

(Irwin, 2003, p. 67)

I am staying "attuned to the aesthetics of unfolding insights."

(Irwin, 2003, p. 68)

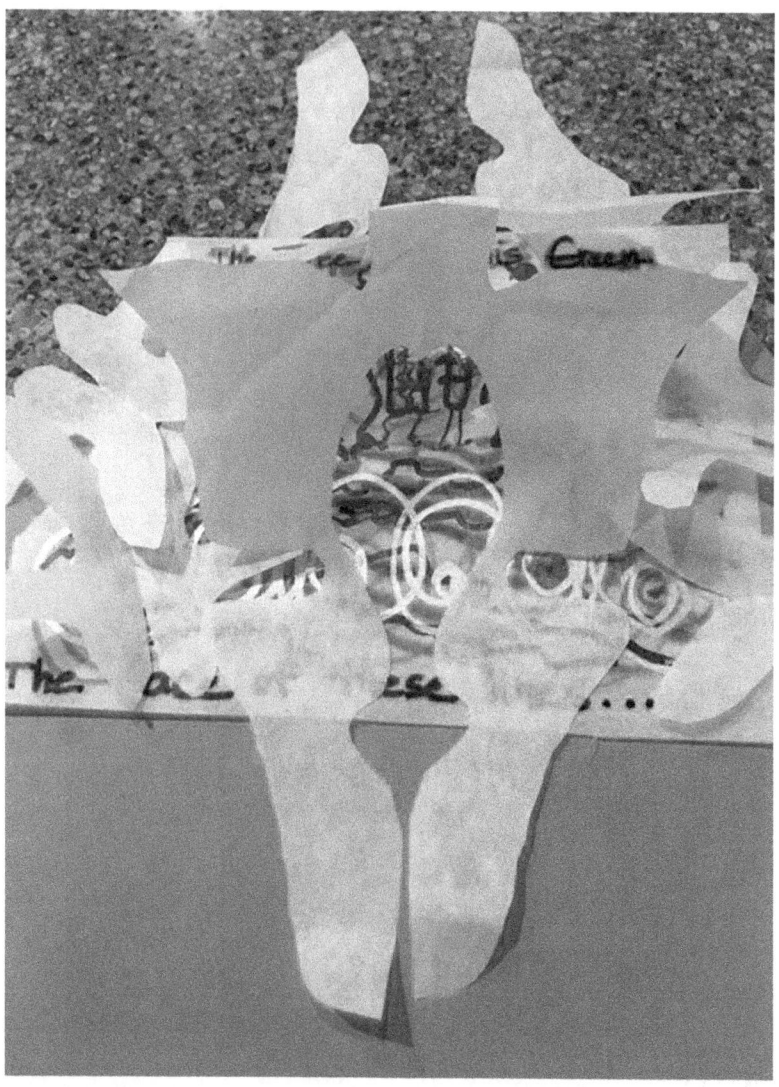

Figure 5.6

"The whole canvas vibrates white. The movement becomes pulsing" (Manning & Massumi, 2014, p. 77). "Between transparency and the opacity of white there exists an indirect number of degrees of cloudiness."

(Delueze, 1986, p. 93 in Manning & Massumi, 2014, p. 77)

Figure 5.7

Knowledge is created through these alternative forms of inquiry, and as educators . . . sensing ourselves to these ways of knowing opens up deeper understanding towards our day-to-day negotiations with others and ourselves.

(Irwin, 2003, p. 72)

I need my intercessors to express myself, and they could never express themselves without me: we are always at work even when it is not obvious.

(Manning & Massumi, 2014, p. 65)

Fictive or real, animate or inanimate, our intercessors must be created. They come in series.

(Deleuze, 1995, p. 125 in Manning & Massumi, 2014, p. 64)

The intercessor is a complex singularity that activates a process, a force that acts as a differential within an ongoing movement of thought.

(Manning & Massumi, 2014, p. 65)

Figure 5.8

Figure 5.9

Figure 5.10

Painting is not about seeing. It is felt. It touches, it moves, it resonates.
(Irwin, 2003, p. 64)

I affirmed the unfolding process and worked through the surprises that
might have seemed to be errors, only to discover a richer and more
complex resolution.
(Irwin, 2003, p. 70)

Figure 5.11

Figure 5.12

Figure 5.13

I was attuned to the process and the image being created.
(Irwin, 2003, p. 70)

Aesthetics stems from the Greek root word width, and especially the very aisthanomai, to mean feeling through a heuristic act of perception.
(Irwin, 2003, p. 71)

This entire process "required a surrendering on my part."
(Irwin, 2003, p. 69)

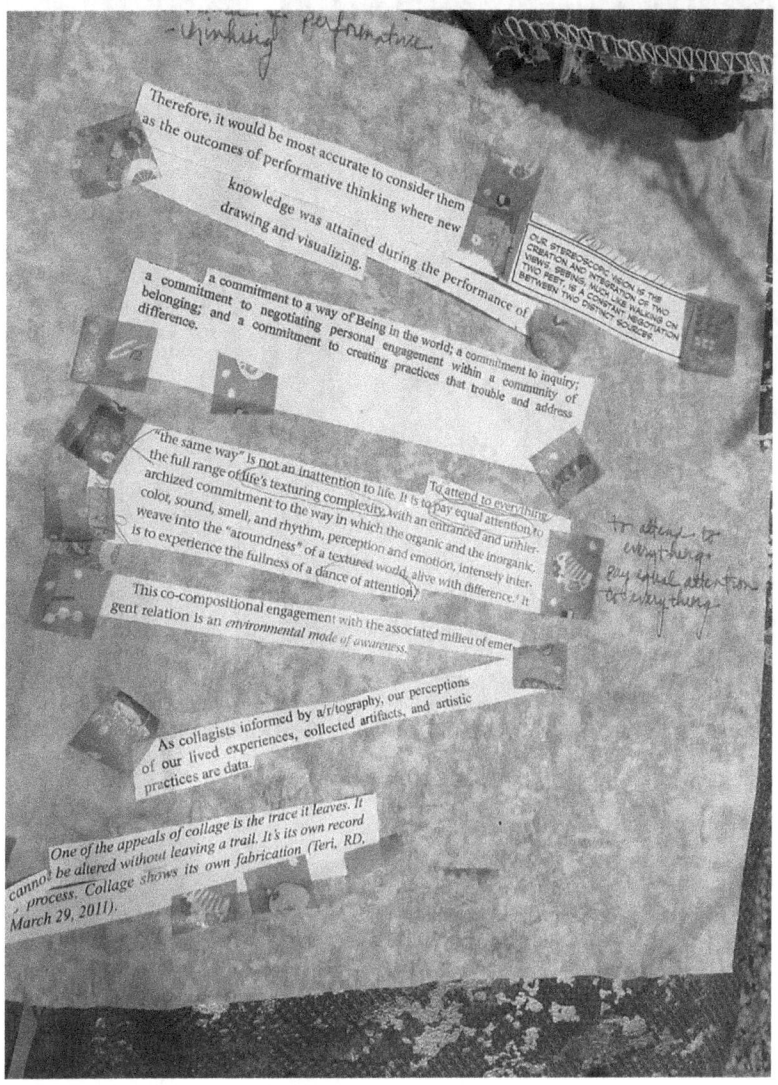

Figure 5.14

Figure 5.15

As artists see attentively and create layers upon layers upon layers of visual images upon the other, they are attuned to the relationship between them and the work of art.

(Irwin, 2003, p. 68)

Color . . . . It is not part of the beginning decision . . . the lines and the colors come together . . . . Starts with the same color as the last, until the painting asks for another color.

(Manning & Massumi, 2014, p. 74)

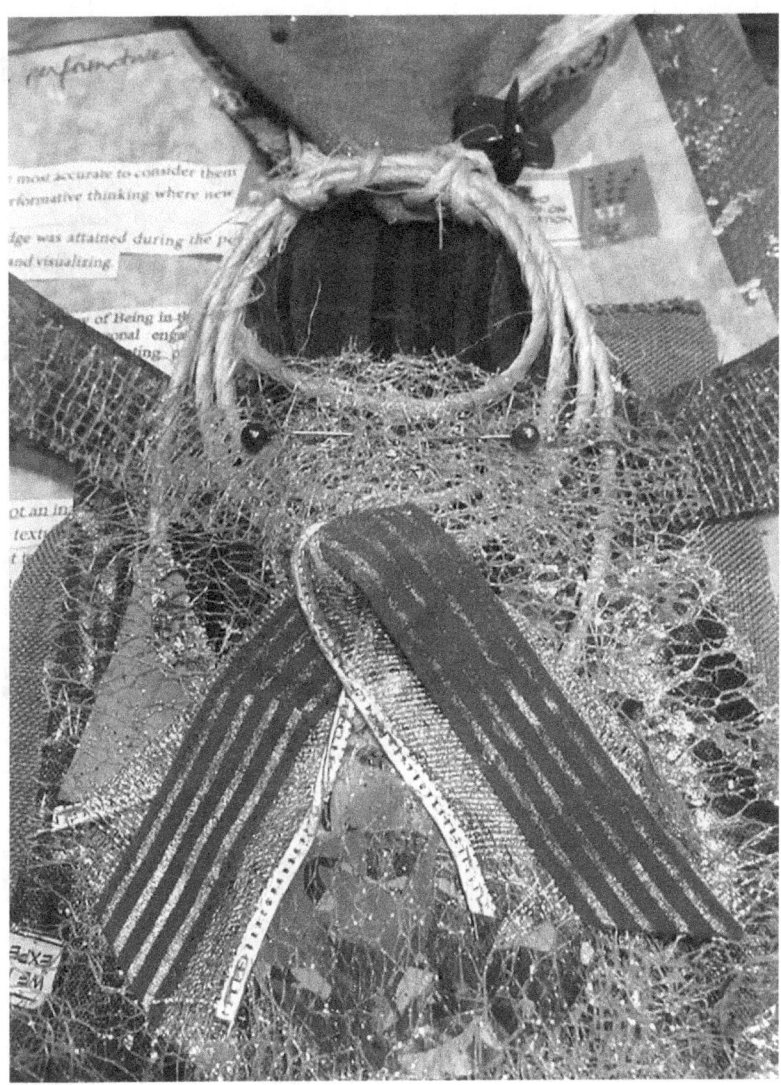

Figure 5.16

Figure 5.17

Figure 5.18

Figure 5.19

Figure 5.20

Figure 5.21

*The painting process is its own thinking-feeling subject, moving choos-
ily across material operations.*

(*Manning & Massumi, 2014, p. 80*)

*Through the exercise of its power to choose itself, the painting process
becomes a formative force of thought.*

(*Manning & Massumi, 2014, p. 79*)

I return to Szekely again when he states; "Play releases the enormous tension and pent-up emotions that easily build upon a highly regulated school
environment" (2015, p. 13). While this speaks towards K-12 education,

Figure 5.22

I believe this notion should be revived in higher education, in particular, to free imagination, make space for wonder and possibility, and incorporate the inclusion of multiple literacies. In addition, permission to slow down and spend time in this serious play affords opportunities for insights. Throughout the course, TWK relaxed into the intensive process of play and discovered that the identity of artist goes far beyond the capacity to render well.

> *I've really tried to suppress the vulnerability, confusion and chaos in between each moment of performing process as a researcher, teacher, and artist over the past 10 weeks. Ok. I will say it. I'm an artist. It isn't easy for me to say. I am struggling to embrace this identity*

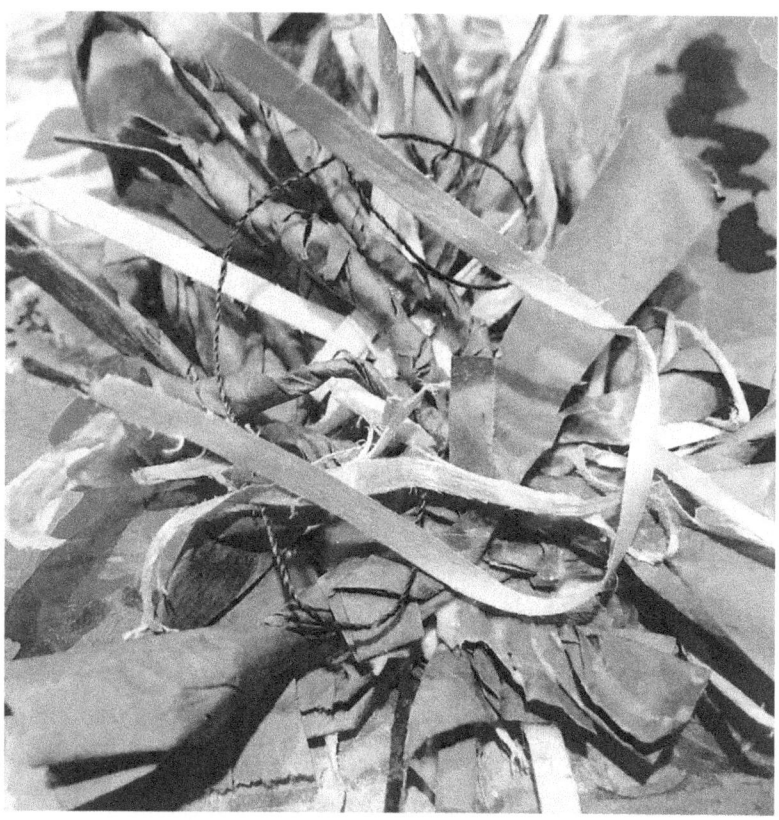

Figure 5.23

*and part of myself. But I am saying it with a very small squeak. This
new world has opened very wide to me and I'm trying to bend and
make my "things of power" bend towards my ability to use them in
my research, teaching, artistic practice, and play. Through marching squiggly stick people, to the Red Lady and Pink Lady, then the
surprise of unknown openings with paint for the first time, and finally
woman, who becomes . . .*

An artist

*Mazzei (2014) says: "Diffraction . . . [is] a physical phenomenon . . .
when ocean waves pass through an opening or obstruction and are*

Figure 5.24

Figure 5.25

generally spread differently than they would be otherwise . . ." (p. 742).
It is a physical and spiritual awakening to be drawn to play and create
FIRST before writing or speaking. In fact, I was wandering and scram-
bling in-between each interaction of my ladies and then finally She's
There. The unknown openings washed me into a doorway of playing
with white paint with no plan at all. It just came to be as I sat down
to play.

I am becoming much more attuned to being accessible to difference
and taking time to absorb what might be ignored. I'm actually gather-
ing insight. I take notes more often. I map. I find myself being observa-
tional as a way of thinking, and then appreciating that I took the time
to document that moment or idea.

References

Barad, K. (2007). *Meeting the universe halfway: Quantum physics and the entanglement of matter and meaning.* Duke University Press.

Deleuze, G. (1986). *Cinema: The time-image* (Vol. 2). University of Minnesota Press.

Deleuze, G. (1995). *Negotiations, 1972–1990.* Columbia University Press.

Holbrook, T., & Pourchier, N. M. (2014). Collage as analysis: Remixing in the crisis of doubt. *Qualitative Inquiry, 20*(6), 754–763.

Irwin, R. L. (2003, Fall). Toward an aesthetic of unfolding in/sight through curriculum. *Journal of Canadian Association for Curriculum Studies, 1*(2), 63–78. https://doi.org/10.1057/9781137015839.0013

Jackson, A. Y., & Mazzei, L. A. (2013). Plugging one text into another. *Qualitative Inquiry, 19*(4), 261–271. https://doi.org/10.1177/1077800412471510

Le Freitas, E. (2017). Karen Barad's quantum ontology and posthuman ethics: Rethinking the concept of relationality. *Qualitative Inquiry, 23*(9), 741–748.

Manning, E., & Massumi, B. (2014). *Thought in the act: Passages in the ecology of experience.* University of Minnesota Press.

Mazzei, L. A. (2014). Beyond an easy sense: A diffractive analysis. *Qualitative Inquiry, 20*(6), 742–746.

St. Pierre, E. A. (2016). Untraining educational researchers. *Research in Education, 96*(1), 6–11.

St. Pierre, E. A. (2017). Deleuze and Guattari's language for new empirical inquiry. *Educational Philosophy and Theory, 49*(11), 1080–1089.

Szekely, G. (2015). *Play and creativity in art teaching.* Routledge.

6 Re/presentation

Chapter 6 follows Brooke's (pseudonym) intra-actions between theoretical readings and material exploration that were assigned as homework throughout the duration of the course. Traditionally a digital artist, Brooke was invested in the idea of representation. She had already completed an MFA and was teaching undergraduate courses in 2D design. Realizing and appreciating the complexities of teaching, she elected to return to school for a deeper dive into art education and research and to earn her PhD. In doing so, she began to explore her own long-time artistic process through the a/r/t/ographical lenses when she enrolled in the Artistic Research course. Historically, Brooke's work was clean-lined, organized, and very product-oriented. Her work explored the literal representation of an idea or topic of exploration. She had developed a method of working and thinking that was successful and comfortable for her. Asking her to move beyond her familiar methods of artful expression to disrupt habit and think differently, Brooke elected to explore theory through the use of childhood toys. I was tickled by this and wondered if by selecting toys, she made a space for play much the way children engage in the moment with play. Szekely, (2015) offers the following insight regarding engagement with toys; "Each toy has a story, suggests an adventure or fantasy. Toys are special vehicles for the imagination" (p. 88). Toys express agency and invite intra-action while also disarming the seriousness we often assign to graduate work and research. Toys and play anchor us in the moment, drawing us into the process. Remaining fully present in the moment enabled Brooke's attention to shift and become more deeply aware how her thinking unfolds. Her attention to tension enabled her to uncover deeper ways of interrogating ideas and discovering new knowledge.

Below are selected writings from her weekly blog, accompanied by the voices of the authors in the readings, and my own responses and interactions with her intra-actions. The selections were chosen as they specifically illuminate the transformation in her thinking as she moved through the

DOI: 10.4324/9780367823818-6

assignments. The tensions between notions surrounding traditional representation at odds with the ambiguity of the unknown invited Brooke to think differently about process and undo some of her long-held beliefs about knowledge and research. As in Chapter 5, play becomes a vehicle for exploration here as well. What follows offers a window into her explorations and philosophical contemplations. The narrative is recreated as a conversational script accompanied by the visual voices of the toys she engaged with. All scripted content is taken directly from Brooke's blog while the remaining text discusses and comments on the unfolding play.

Week 2

BROOKE: *So, I will start off by saying that I have a lot of questions about representation. I am trained as an artist, not necessarily a researcher, so as an artist I am used to representing things. However, I think that the term "representation" means something different in terms of research. But I am still having trouble wrapping my head around it. Isn't everything a representation?? Like in the blogging article, the author talks about the materiality of our practices and documenting experiential encounters . . . isn't that an attempt to represent our findings?*

MICHELKEVICIUS: *"Is this a piece of art? Or is it the outcome of the research process? Who made it? Therefore, its status is considered to be ambivalent. From the perspective of the intention (to present the most iconic photograph), it could be called a product of curatorial research; however, its aesthetic form refers to an art piece (Michelkevicius, 2012, p. 123)."*

" . . . the performance practices she writes about do not illustrate the concept of immanence in a representational fashion, so much as they try to perform immanence; it is less a matter of trying to show what immanence looks like and more a matter of figuring out how to be inside it and then seeing what comes out of that experience to immanence itself (Springgay, 2017, p. 275)."

BROOKE: *They are both not necessarily about the end product, although they do produce something in the end. I think that many perceive research and art as being about that end product because that is what they normally experience/see/read . . . but the idea of making the experience or the "becoming" more important is interesting to me and I can see why it is troubling traditional notions of research.*

AMY: Hi Brooke!

I love reading your train of thought and following along with your wonderings. I love this; you say, "Art making and research are both about the experience and figuring out how to be inside the process and seeing what comes out of it. They are both not necessarily about the end product, although they do produce something in the end." There is a lot of juice to this. What is happening in that process that it actually produces?

Agency and power, I'd love to hear/see a little more of what that means for you . . .

Your image is very thought provoking, the????? in all the (in)between spaces of folding and moving is reminiscent of the and, and, and, and . . .

How do you see/sense these three ideas of representation, agency, and intent "mattering" inside of process?"

Early in her experiences, Brooke struggled with focusing on process over product. Representation meant a finished object, and that finished product was privileged over process. There was a correct path to reach an acceptable representation of an idea through material and objects. Art produces a predictable product when step-by-step instructions are followed and replicated. In K12 art education, we often hear this referred to as cookie-cutter lessons. These are lessons that are highly scripted and where, ultimately, the lesson is about following directions rather than creative problem solving or critical thinking. How many of us have strolled down the hall of an elementary school to see beautiful bulletin boards filled with colorful artwork that largely all looks the same? Steadfast methods, while providing artists and researchers a reliable framework, struggles to produce surprise. There is nothing necessarily wrong with that path. The tenets of the Artistic Research course lived in a postqualitative space.

> A post qualitative study cannot and does not begin with any social science methodology, including qualitative methodology, but, rather, with the onto-epistemological arrangement and concepts of poststructuralism and its descriptions of key philosophical concepts such as ontology, epistemology, human being, rationality, truth, discourse, language, freedom, and so on.
>
> (St. Pierre, 2021, p. 1)

It was with this idea in mind that Brooke began her explorations.

Week 3

BROOKE: *TENSION RESPONSE*
   I guess I don't feel tension with the reading itself. I agree with eve-
   rything the authors are saying. The tension for me lies in this act of

intra-action with the text. I feel like I am trying to make it look like what I think is the "correct" way to do it. And it doesn't feel genuine . . . it feels forced. I understand that the assignment description said that there is no right or wrong way to do it . . . but I guess I just feel I am on this quest to do it "correctly", the way that others want me to do it, and not how I would genuinely feel compelled to do it.

The phrase that I keep coming back to is "seeing double". It really stuck out to me because it is about keeping an open mind and different perspectives. That combined with the parts about illumination inspired me to work with light.

While I wanted to refrain from having a "finished product" (but isn't marking the text a "finished product"?) as a part of this assignment, I felt compelled to do something more with the text. I took a trip to Target just to look at objects and materials, seeing if anything resonated with me. I walked down the game aisle and saw the Lite-Brite.

Exploring notions of resilience, illumination, and to some degree improvisation, she allowed herself to playfully enter and intra-act with theory through interaction with unexpected material. Not only was she disrupting traditional notions of text but she was also colliding with the world of play, re-conceptualizing, and repurposing a child's toy to explore theoretical concepts. Her anxiety about *doing this correctly* suggests a need or perhaps reliance on a prescribed method to assure an intended outcome. Without a defined road map, Brooke had to rely on her own sense of wonder. Perhaps, using toys the way that children do with a fearless wonder and ability to take risks without concern allowed her to likewise explore processes and dense theoretical concepts without worry. As she worked through many of her tensions, she allowed herself to take cues and inspirations from specific words that were inspiring or empowering to her in some way.

I wanted to explore the relationship between the things I found interesting in the text and how it "illuminated" my thinking. I first wrote "SEE" in the pegs two times (literal take on "seeing double"). Then I thought about how language is giving too much agency. When we see the letters, we find the word "SEE" and don't think much of it afterward. The language clouds our perception of other elements. So I took the pegs and altered their position in space, still keeping some semblance of the original word. Keep in mind that this doesn't show the effects of lighting, which I can show more in class since our room can get very dark.:)

Figure 6.1 Lite-bright emphasis on word

AMY: **Brooke!!!!**

 I LOVE what you did with the Lite-Brite!!!!

 OK, let's back up here . . . I am intrigued with the tension you experience about "doing it right". I wonder where that comes from? Can you follow that feeling to some time/space/place? How might that internal expectation seep into other areas of thinking and doing?

 HOW you intra-act with the text is up to you (back to LOVING the Lite-Brite exploration and your explanation of it). You highlighted several phrases from the reading:

 —*strengths can become traps-uniqueness of perspectives = kaleido-scopic views-think about seeing-seeing double-no single or "correct" view-trapped within borders of vision-what was outside is replicated within*

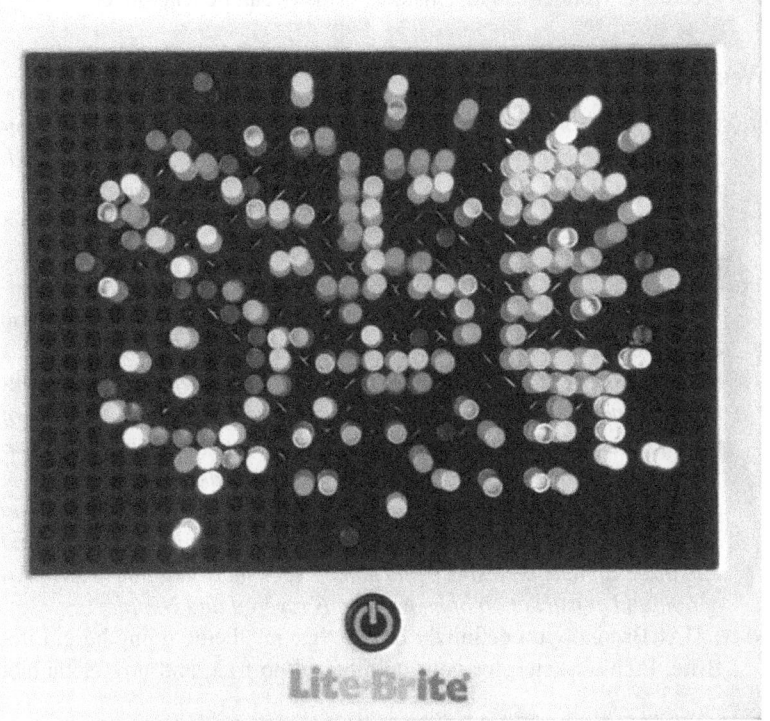

Figure 6.2 Lite-bright emphasis on image and light

HOW does visually intra-acting with the theoretical constructs behind these statements illuminate and expand understanding and HOW it operates in your own thinking? In sinking into something, spending time, dialoguing through multiple literacies, something happens. (Well, hopefully). I can't wait to see this in class! I LOVE how the same points of light can be rearranged so in one formation the "meaning" see, (privileging the word while the actual color and light is secondary), but in another, we "see" light. By working with this, you've literally illuminated different ways of knowing and encountering.

Some weeks there may not be tensions in the content of the reading and that's fine! Other weeks there likely will be. (Remind me to talk with you more about your questions about representation . . . I've been simmering with that all week). While you agree with all of it thus far, sometimes getting to know a part of it more intimately produces

something new and unexpected. Tension doesn't have to be about disa-
greement. Wonder can be tension, curiosity can be tension, etc.

Week 4

SHOTTER: *"A performative understanding of scientific practices, for exam-*
ple, takes account of the fact that knowing does not come from stand-
ing at a distance and representing but rather from a direct material
engagement with the world (Barad, 2007, p. 49 as quoted in Shotter,
2014, p. 305)."

BROOKE: *I picked out the quote that is at the top of this post because it reso-*
nated with how I am making sense of research and how I make sense
of the world around me. We are never just simply outside observers as
researchers. We are not just recording what is happening and present-
ing our findings. We are a part of the research. Research is a part of
who we are. Everything is crossing and becoming a part of each other.
This reminded me of Scrabble. So, I decided to take a few words from
the readings that I thought were powerful in regard to this topic and
have them intra-act on the Scrabble board. But I wanted to bring it out
into three-dimensional space and make it look more sculptural as intra-
action and construction of knowledge is layered and complex.

AMY: HA! Brooke, you definitely have a "game" theme going here. Lite-
Brite, Etch-a-Sketch for your ontological portrait, and now Scrabble!

Figure 6.3 Word play with scrabble

I am so curious and interested in these choices, it's as if these objects become the "medium" (think fortune teller) (in)be/tween you and the theory. As I sink into looking at your response, my own impulse is to continue exploring the letters which make the words which impart some sort of meaning (constructed) from our experiences, and turn letters in different directions to further complicate and deconstruct the power to the language, making it progressively more visual. The essence of the original "ordering" there, but now completely transformed. HOW did thinking through these ideas in this manner extend/expand or transform your thinking about the quote you shared? I also noted the word "representing" again . . . :)

Week 5

BROOKE: I *think I am finally realizing why the concepts of this course seem so foreign to me (although exciting, too). As I mentioned in my artist talk last week, going through my MFA program, students start off with very broad concepts and experiment with different media and processes, then they have to home in on them and get very specific. This was necessary because graduating students had to present a final exhibition and a 50+ page written thesis. So in the MFA program, it was essential that I know exactly what I was doing and why. Everything, down to the colors that I used, the size of my work, even the lighting, had to be meticulously calculated, assessed, and explained in written and verbal form. Of course, I understand the importance of this, but now that I am in a different context—art education and research—I am having to think about things in the opposite way. How I have been trained through my education as a studio artist has been to get very specific with my work, my processes, and my ideas but in my current context, it is to get broad again and explore different processes and ways of knowing. It sounds like a relatively simple task, in theory, but when I have gone through the past seven years getting super specific, it is difficult to get out of that mindset! It is also incredibly freeing, though. I don't need an answer for every action, mark, or representation. At least, not for now. And this is not to say that I don't play in my current work, but it is nice to play without the constraints that I'm used to.*

As I was reading "Thought in the Act" I came across this passage that I want to focus on this week, with the above thoughts in mind:

MANNING & MASSUMI: " *. . . play being the minor tendency contained by every instituted structure, whose unleashing softens or disables postural default settings. A degree of play creates the potential for the* *emergence of the new, not in the frontal assault against structure but at the edges and in its pores (Manning & Massumi, 2014, p. 99)."*

BROOKE: *That passage resonated with me, above all others, because play is something I am struggling with being in academics. Play is risky in the context of being a student. Students have to produce for a grade, for a dissertation, for a future job. Play doesn't always produce a tangible/readable/viewable product, but it is also necessary. New ideas/ thoughts/projects come out of play. With that passage, I am thinking specifically about my practice as an artist, a researcher, and a teacher. I see it as aspirations with each part of the a/r/tographer identity.*

AMY: **Brooke!!!!!**

Your emerging awareness is bringing you to a threshold! That (in) between of a collision of—either/or- now inviting the -and/both-. The MM quote you selected had me jumping out of my chair with excitement!!!!

I love the Jenga exploration. It is a fabulous visual metaphor for the thoughts you articulate so beautifully. I am intrigued by the differences in how focus is emphasized in the two programs. There is nothing wrong with either, and again, you invite both modes to the table to dialog. I am wondering if you might apply this to your current aesthetic? You say, "There are constraints in that I don't want to lose my identity as an artist, that includes my aesthetic. I see my aesthetic as something that I have worked really hard for." Absolutely. It is yours and it is strong and will continue to operate from within. It's a part of you so how can you lose it? BUT, what if you "trouble its structure"? Play is not confining nor defining, it is simply a wondering. What if rather than losing your sense of identity it be/comes something more? Sit in the tensions of this and pay attention to what that produces . . .

Week 6

JACKSON & MAZZEI: *"Plugging in to produce something new is a constant, continuous process of making and unmaking. An assemblage isn't a thing-it is the process of making and unmaking the thing. It is the process of arranging, organizing, fitting together. So to see it at work, we have to ask not only how things are connected, but also what territory is claimed in that connection (Jackson & Mazzei, 2013, p. 1)."*

BROOKE: *Speaking from an artist identity, when I am making a piece, I am adding and subtracting imagery. When thinking from the researcher's identity, I go through the same processes, but it is less visual and more about language. When I am writing about research, I am thinking a lot*

Figure 6.4 Jenga as a metaphor for risk taking during play

Figure 6.5

Figure 6.6

about the words I use and how I can express language in a way that flows and makes sense and sounds maybe more poetic and soft and accessible. So, I will continually switch out words for synonyms that sound better to me. But sometimes, in doing this, I have to reevaluate the whole sentence . . . and then sometimes the whole paragraph. Thinking more about expressing things visually and through language/ writing, I do sort of a stream-of-consciousness approach and start writing/drawing whatever comes to mind. Then, through the process of editing, I start to make more conscious decisions and contemplate more on the meaning behind each word/image.

In many ways, I guess, language and images work in the same way. There are, of course, differences, but reflecting on it through these readings made me look at how I am seeing each component. When I "plug in" text/language how much different is it than when I "plug in" images? Just like in language, image has limits. There is only so much one can say in a work of art just as in how one uses words . . . but they can both still say a lot. And, oftentimes, the importance doesn't lie as much in <u>what</u> we say or <u>what</u> images we chose, but the <u>how</u>. How am I using language? How am I using images?

JACKSON & MAZZEI: *"Matter, both human and nonhuman, becomes vibrant in an assemblage; objects take on 'thing-power'. That is, objects become things when they become energetic and make things happen. Agency, in this assemblage, is spatially distributed among vibrant matter, rather than traced to a single source or marked off by a particular boundary (Jackson & Mazzei, 2016, p. 95)."*

BROOKE: *I can see how I am "plugging" the Agentic Assemblage article with the Plugging text. By thinking about both at the same time and how they are similar and different, I can visualize an assemblage starting to take place. It is making me think sculpturally. Data as tangible objects that are distributed widely with no boundaries or a direct, single source of knowledge.*

So I kept thinking about this overall, broad, unbounded emphasis. Paying attention to the subtleties and the "plugging in" of one text/ image to another. I thought about the Lite-Brite and wanted to go

Figure 6.7 Revisiting lite bright with a broader lens

*back to that. With the Lite-Brite, I can literally "plug in" a "concept".
I thought about each color being a different idea/concept. I wanted to
visually show the act of plugging in and revising and letting things be
broad without one single emphasis point . . . While it certainly would
have been easier to have a plan to start off with and calculate where
each peg would go, that is not how research works. If it did work that
way, there would most certainly be biases and an inauthentic result.
Below are a few pictures of that process.*

AMY: I LOVE THIS—"I started to begin visualizing data as an assemblage
after reading it." This is so powerful!

You also say, "How am I using language? How am I using images?"
YES! Then we enter the world of doings!!! Verbs. Who knew verbs
were so cool? You make some really good points about the research
process, having a plan versus not having a plan. Hence, methodologies. So more post(y) researchers argue that qualitative research has
become far too predetermined, while having a protocol somehow lends
replicability and validity to outcomes, if we know ahead of time what
we are looking for, we are sure to find it. So I LOVE what you did with
the Lite-Brite, allowing the process to illuminate and unfold as you
all (meaning the pegs, the surface, the light, etc.), participated in the
process, so agency is plugged in here as well.:) I am curious, do you
notice shifts in your thinking and perception working with these heavy
theories in this manner?

Week 8

IRWIN: *"The slash within in/sight embellishes the unperceived held within
the perceived. It highlights the dialectic between insight and out of*

Figure 6.8 Folds and crumples as in-between or lost spaces

sight. It attends to the tacit knowledge that remains unspoken yet recognizable, honoring the mysterious and the ineffable. The aesthetic of unfolding appreciates the awkward spaces between chaos and order, complexity and simplicity, certainty and uncertainty, and cherishes the in-between space that values attunement. Unfolding appreciates creativity set within emergent, dynamic, and awkward spaces that, given time and space, develops into complex aesthetic creations (Irwin, 2003, p. 77)."

BROOKE: *As an artist and researcher, I am ok with order; but it is the chaos part that I have trouble incorporating. I do know, though, that this is where invention occurs and cultivates the most interesting doings/learnings . . . As I was reading this week, I kept thinking about origami. There is a great documentary out there called "Between the Folds" that I had been showing to my 3D design students for years before it got pulled from Netflix. But the documentary talks about the history of origami and how it lingers between chaos and order in the most remarkable ways. There are artists out there who make super mathematical, complex pieces and some who make very organic, free form shapes. It is very process oriented, with some origami pieces taking months to make . . . While not traditional origami, I wanted to just work with paper and linger between the known and the unknown, complexity and simplicity, and, most importantly for me, chaos and structure. I used regular lined notebook paper because the lines provided me with some structure but I didn't think about what they would look like afterward. I just went with my gut and what I know of origami. The picture below is of the pieces of paper after they have been unfolded. I think this is more of a metaphor for the in-between spaces of thought, as a teacher, researcher, and artist. The fact that this is notebook paper is especially symbolic. Many of us have used notebook paper since the beginning of our educational experiences to collect information. The notebook paper is lined and tells us where to write notes. It is providing us a structure for our thoughts. Normally, when paper is crumpled up, that signifies trash or knowledge that was not of value or is no longer needed. When one goes through ideas, they might jot them down on paper and discard the ones they don't need; but really, it is all important. Those ideas that we are so eager to discard are valuable, they may be even more valuable than the idea you decided to go with in the end. The crumpled/folded/discarded paper, for me, represents those in-between and often lost spaces.*

AMY: **Brooke,**

This is a fabulous statement: "There are complexities to life that may require that we hold one side or the other, but learning really relies on the dynamic in-between spaces. Learning relies on the realization that we don't know everything (gasp!). We can all set ourselves and our students up for quality learning experiences by surrendering to the fact that there is always more to learn and always new things that will and should challenge our previously held notions." YES! A continuous/contiguous journey of *becoming* discovery!

Brooke, the manner in which you explore these concepts each week is truly profound and fascinating. I sometimes wonder if you imagine you must let go of all you've known to embrace these other concepts. I hope you are sensing that that is not the case. All you are and have been, you bring to the present moment as you contemplate the newness of the unknown. You weave them together in that in/between with such honest exploration!

Week 9

MANNING & MASSUMI: *"the viewer does not have the axis"*
"the painter does not have the axis"
"Their movement is not governed by any direct application of principles of composition. It continues to emerge, in a manner that can still be called machinic-if that word can be used for a form-coming that occurs in a suspension of authorial will and spectatorial perception, in relation to an outside of those axes, immanent to the painting process (Manning & Massumi, 2014, p. 60)."

BROOKE: *While reading and writing this week, I felt some frustration with these ideas. I don't think it really has anything to do with the readings, though. I think the frustration comes from a place deep inside that is still trying to show art as valid through intellect. Painting from intuition or painting from feeling is not something I am totally comfortable with. The idea is more important than the process for me. I don't know if it will always be that way, but right now at this point in my life, that is what drives me and that is what I am drawn to. That is also how I make sense of art, though intellect. Which is what I think the readings/videos for this week are saying not to do haha! I will think more about this . . . it has been something that has been on my mind a lot throughout the Art Education program. For this week, I decided that, because of my frustrations, I would just make something based on emotion and intuition. I did decide to do it digitally because I didn't have adequate art supplies at home tonight. But below is a video (sped up) of this process. I didn't have a plan in mind when I started and there wasn't a concept behind it. It was just me chasing colors and shapes based on what I felt. I was trying to push concept to the side, but, during this process, I was wondering if that was even entirely possible. But I was trying to keep things moving, activated, and let the process of making the work drive my thinking . . . trying to draw in a way that wasn't totally under my control/preconceived ideas.*

AMY: You say, "That is also how I make sense of art, though intellect. Which is what I think the readings/videos for this week are saying not to do haha! "Maybe it's BOTH/And????

Then you say "I will think more about this . . ." . . . an intellectual process. What if you feel more about it?;) You explore that with your digital exploration . . . . What did it produce?

Week 10

By week 10, the students in the class were beginning to synthesize these experiences, and very individualized outcomes were unfolding for each one. I asked them to comment on the following question: How does the idea of diffraction blanket much of what has been unfolding in these past 10 weeks?

BROOKE: *This idea of diffraction is most obvious in the encouragement of working in new materials. It is a lot like "plugging in" different theories when looking at data. We all have our unique visual sensibilities, but when we work with those sensibilities in a way that is disruptive to our knowing, it may change how we view/interpret/intra-act with those ideas. In some ways, we are noticing the differences we feel or the differences we show when working with a new material. I have not been working digitally for a majority of this semester for this class and I have noticed differences in the content I am gravitating towards. I am thinking more sculpturally with found objects (i.e. games). My 2-dimensional work is often capturing a moment in time, whereas my 3-dimensional work focuses more on the process and the action/intra-action with those objects. Although I am working with familiar objects in both my 2D and 3D work, I am creating relationships that are new by working outside of my habitual art making process. By working outside of my comfort zone, I am allowing for the space to create new knowledge. However, I will say that this can still happen when working within a medium we are comfortable with, it would just require that the maker intentionally test that medium with new processes. So,*

I don't think that I necessarily needed to abandon making things digitally. I could have worked with something like a new app, for instance. But it definitely speeds up and forces the process of thinking diffractively when one completely goes in the other direction. Ultimately, I am happy that I chose to work with the materials I worked with for this class, as there is only so much time to allow the process to take shape and challenge habitual thinking.

I then asked them to select ONE sentence or passage from the reading that in some way FEELS (FELT?) diffractive. Discuss from the three perspectives of the artist, researcher, and teacher, individually or in combination.

JACKSON: *"The outside of method, then, is comprised of relations between a dance of forces in an emergent, non stratified space of resistance and a nonplace of mutation, where 'suddenly, things are no longer perceived or propositions articulated in the same way' (Deleuze, 1988, p. 87). To be on the outside of method, then is not an exteriority . . . The outside is the transformation itself (Jackson, 2017, p. 2)."*

BROOKE: *This passage feels diffractive because it is disruptive to how I have been taught, in some classes, to think about research. I have been told that I must start with my research question then choose my methodology and stick with that method throughout my data analysis. But to think outside of method is somewhat foreign for me in that it is an area of the unfamiliar and unstructured. Beyond the lens of the researcher, I am more comfortable with this area as an artist. Artists are encouraged to think in new ways and outside of the normative processes. So, in that way, I can more easily relate it to the perspective of a researcher. I read this passage before I installed my show on Saturday and I decided to film the process of installing. However, I wanted to show the process in reverse because viewing it in that way, things were no longer perceived in the same way. It was much like drawing a representation of something upside down. In doing so, we notice shapes and lines that we didn't notice by perceiving things the normal way. When I was looking at things in reverse, I thought it was very metaphorical of diffractive thinking. If you view it with the sound on, the voice and noises sound like computer glitches or something . . . but it is not the language that is important. The importance of this video in terms of diffraction for me, is that I am in the process of undoing everything I had planned. Un-measuring, un-hanging, un-planning. Starting at the end and working my way back to the beginning, emerging/ending at a place when I didn't have a plan. The video in its normal linear orientation from start to finish is privileging the knowing, but doing it in reverse privileges the unknown, the moment of possibilities. Below is my reverse installation process video, my intra-action for the week.*

They were then asked to comment on the relationship of the screenplay with the video. Where do you see the "mattering" of multiple theories?

BROOKE: *I am in the beginning stages of my screenplay/video and, I'll be honest, I have never written a screenplay before. However, as I am looking through my videos and mentally "writing" the screenplay, I am thinking about how I might be able to take the footage and use it in a new way. The screenplay is maybe more ambitious than the videos . . . or maybe it is that the videos are not saying everything in a way that I feel encompasses the things that I have learned. As I am going back and viewing all the videos, I want to see them in a new way. I guess I want to do something with drawing or painting the intra-actions, providing another layer. So, the final video might end up being about my intra-action with my intra-actions throughout the semester. I am thinking about the mattering of multiple theories in the same way I think about the mattering of multiple art media. As I stated above, the use or "plugging in" of different materials may produce different results.*

AMY: WOWOWOW. Just wow, Brooke! Your consistently deep contemplation and surrender to this process is incredibly profound. It's almost as if conceptual thinking has become its own agentic presence within and in-between your prior knowing. You say, "I find myself thinking about the text visually and the visuals in terms of text." I have often thought this as well. I found a similar realization awakening as I created my own movie and screenplay. Keep pushing and exploring, Brooke!!!!

Week 11: Composing a Documentary

BROOKE: *Below is my teaser trailer for my documentary. I chose to do an animation because, as I was looking through my footage, I felt compelled to create an intra-action of my intra-actions. I am not great with animation (as you can probably tell haha), but I wanted to try it. I think I may mix in this animated work with my real-life footage for my final documentary (partly because this takes sooo long to do). When doing the animation, I had to work slow (reeeeaaally slow) and obviously frame by frame. This reminded me of what I was going through as I was reading some of the text. With the readings and the animations, I was learning as I went and I often had to backtrack and see what I read/did before in order to go on to the next step. With animation, you have to look at movement and placement of shapes closely to try to accurately mimic movement and continually be reflective of past placements. With reading, especially for the text in this class, you have to look closely at words, their context, and continuously reflect on your ontology, epistemology, and axiology to determine whether or not you agree with the author(s) and/or where you stand in terms of the writers system of beliefs. Reading and animation are a lot alike in these ways. They are both slow and require continual reflection. Taking my IRL footage and using it for the animations is encouraging me to take the most important pieces out of my IRL footage and simplify my thought process somewhat. This, like my previous intra-actions, has made me dive into meaning in a way that I just couldn't access through text alone. I learn visually and in working with my hands (by working with painting materials, sculptural objects, the computer, etc). Specifically in working frame by frame, it is a way that I am able to organize and compartmentalize my thoughts. At the same time, my goal for the rest of the documentary will be to make sure to not over-simplify things and keep the nuances of my thought process and the uncertainty apparent.*

I see this process of creating the documentary and the screenplay as research in that it is mining from very disparate areas of thought (text and image) to create meaning. I'm thinking back to Jackson and Mazzei's "Plugging one text into another" in particular . . . I find myself thinking about the text visually and the visuals in terms of text. In a way, I see this as plugging in and using the very different attributes of each to inform the other and create a more nuanced and meaningful intra-action with both. Editing is also research because it requires reflection and diffraction. What do I need? What can I take out? Why does what I keep matter more than what I discard? How predictable do I want to make this? How can I pull away from my habitual ways of knowing and doing to try to capture something that is more true and new?

As Brooke speaks to in the foreword, getting messy shifts focus. Playing grants permission to wonder and discover in different ways. Ending with more questions seeds the ground for new and unexpected encounters rooted in the process.

References

Barad, K. (2007). *Meeting the universe halfway: Quantum physics and the entanglement of matter and meaning.* Duke University Press.

Deleuze, G. (1988). *Foucault.* Minneapolis: University of Minnesota Press.

Irwin, R. (2003). Toward an aesthetic of unfolding in/sights through curriculum. *Journal of the Canadian Association for Curriculum Studies, 1*(2).

Jackson, A. Y. (2017). Thinking without method. *Qualitative Inquiry, 23*(9).

Jackson, A. Y., & Mazzei, L. A. (2013). Plugging one text into another. *Thinking with Theory in Qualitative Research, 19*(4).

Jackson, A. Y., & Mazzei, L. A. (2016). Thinking with an agentic assemblage in posthuman inquiry. In C. A. Taylor & C. Hughes (Eds.), *Posthuman research practices in education* (pp. 93–107). Palgrave Macmillan.

Manning, E., & Massumi, B. (2014). *Thought in the act: Passages in the ecology of experience.* University of Minnesota Press.

Michelkevicius, V. (2012). Notes on interdisciplinary methodology of artistic research: Visual thinking, writing and mapping. *ACTA Academiae Artium Vilnensis, 67.*

Shotter, J. (2014). Agential realism, social constructionism, and our living relations to our surroundings: Sensing similarities rather than seeing patterns. *Theory & Psychology, 24*(3).

Springgay, S. (2017). Learning to be affected: Matters of pedagogy in the artists' soup kitchen. *Educational Philosophy and Theory, 49*(3).

St. Pierre, E. A. (2021). Why post qualitative inquiry? *Qualitative Inquiry, 27*(2), 163–166.

Szekely, G. (2015). *Play and creativity in art teaching.* Routledge.

7 Validation of the Visual Voice

Humbled to share that my play, also a chapter of my dissertation, received the National Partners of American Theatre (NAPAT) "outstanding play" award.

(Gina, 1/2022 personal Facebook post)

The semester after the Artistic Research course ended, I was approached by Gina (pseudonym) to do an independent study so that she might continue her work as she prepared for her dissertation proposal. I was excited at this prospect! She knew I had been gathering data in the previous course and she offered that this might provide continued data for my own research beyond the course. We were both excited by the possibilities and mapped out encounters and questions to take a targeted deeper dive, capitalizing on the momentum she had from the course. Since the video intra-actions with theory had been helpful, she expanded the dialogue to include photos and visual artifacts she was using for her personal research. Outside of the formal classroom, she continued to wrestle with her perceptions as an academic and the validity of nonverbal research. This became a subtheme in our conversation as she wrestled with conflicting notions of identity in academe. We met four times throughout the semester to discuss her experiences with reading and intra-acting. She kept up her blog as well as part of the experience which provided additional insights in between our interviews. I recorded and transcribed all of our interviews.

What follows are selections from her blog and our extensive conversations which were transcribed as Gina navigated the moving parts of her dissertation proposal and the legitimacy of nonverbal research methods. There were 200 pages representing over five hours of discussion. It was challenging to edit down so much rich discussion and reflection. I made the decision to highlight excerpts which illuminated her ideas for her research while also giving attention to shifts in her own self-conceptions as a scholar. As in

DOI: 10.4324/9780367823818-7

previous chapters, *Gina's voice is identified through italics,* **citations she is responding to are bold italics,** while my voice, including citations within, is standard text. Her personal blog contemplations are indented. This first blog highlights her experiences as she revisited some of the readings and content from the Artistic Research course.

Blog Contemplation #1

In Holbrook and Pourchier's, Collage as Analysis: Remixing in the Crisis of Doubt Qualitative Inquiry, the author's state, **"Because our collages are exposures and not answers, they are always one of many—multiple and ongoing as long as we continue to inquire"** *(2014, p. 761). I chose to intra-act with this section, as a way to explore ways that collage making could serve purpose and meaning in my dissertation. I have photographs that serve as data for one of the chapters in my dissertation. I felt it was important to incorporate the photographs as part of my data, but I have yet to commit to a means of analysis. When I came across this phrase I felt motivated to intra-act with the ideas it presented. The most significant section of words was* **"collage is exposure, and not answers."** *I realized that the only reason I had considered photographs was that I assumed the photographs would be able to answer the questions I had for that section of my dissertation. I assumed I could just plug in the picture next to the arguments made in the chapter. This phrase does not claim that collage making will reveal the answers you are seeking but that instead it is about the process of making and how that experience exposes other ideas, associations, and information one had not considered. I realized, as I spoke to myself during my intra-action, that through collage making I could interpret the meaning of my experiences in ways that I couldn't just by looking at my field notes (text), and that the end goal is not the finished collage but the journey that brings me there. If I were to use collage making as a means of analyzing these pictures then I would grant the collage permission to make the argument, and it would be something I would have no control over, despite what I assumed the picture would tell me. I feel more confident in working on my mini-proposal and making a better case to why intra-acting with the photographs through collage making is a viable option.*

In her first revisiting of the visual thinking and intra-action, she dives into interactions in service to her dissertation proposal. Gina is digging deep into the intersecting relation of literacies and how they work together to disrupt boundaries of thinking. They are in mutual support of one another, symbiotic in nature as opposed to one literacy being subservient to another. In being able to articulate her process so clearly, she elevates the validity of visual thinking and knowing in her own conceptualizations around research. Gina begins to weave the tool of intra-acting into her thinking and investigative process as she sorts through various methods and conceptual frameworks.

Blog Contemplation #2

As Gina began the process of writing a mini dissertation proposal, she wanted to revisit her first encounters with a specific puppet, the Romero puppet. This puppet was the seed for her dissertation topic and as noted at the start of the chapter, a key element in her research. Thus, revisiting and re-searching her past encounters would be a valuable exercise; looking back in order to move forward. She decided to put intra-action to work in service of this discovery process knowing it might unearth not only memories but something new as well.

One of the new additions to my mini-proposal is opening with a short narrative of my first encounter with the Romero puppet. I begin by examining and dissecting the image in hopes that it would trigger memories of the first moment I saw the puppet back in 2004. I looked at the image and began to narrate what I was seeing. I was trying to connect those ideas to my relationship with the puppet. Looking at the image (really looking at the

Figure 7.1 Photo intra-action becomes me and puppet

image) after so many years was quite inspiring. Of course, I am a different person than I was as a young undergrad. And I have been doing so much research on the puppet and performance, plus I now have a relationship with Pine Forest Theatre, (pseudonym) and still looking at this image again helped me recognize the key components of the image/text and how I interpreted these things as a young, brown, theater kid. I realized there were two key elements. The first one was the word "Central American" in a publication about theater. As a Central American theater kid, I quickly gravitated toward the title because I knew that it was about my culture and me. It was also one of the first times that I could identify a direct connection between theater making (Puppetry) with Central America. Second, there is the image of Monseñor Romero as a large-scale puppet in what seems to be a protest or rally. Romero was already an important political and religious figure in my family who advocated for the poor during the civil war, but to see him as a puppet was exciting! Would I have had the same reaction if it was a picture of Romero as a human being and not a puppet? What about the scale of the puppet? And the way his hands were stretched out like Jesus? Silly things but details I rediscovered during my intra-action. I think back to that young, brown theater kid who was searching for a way to fit into a predominantly white, male field. Being able to SEE MYSELF in this cover was life changing so much I continue to be moved by this image even after so many years. In the end my intra-action becomes a new version of the cover with the words "Pine Forest" replaced by "Me and Puppet":)

What follows are excerpts from our dialog. I focus on her discussion around intra-action with the visual and its impact on how she is approaching her thinking about her research.

Discussion #1

AMY: You've been busy.

GINA: *I think stuff is happening now that I understand. I got back into the groove of it. I'm starting to use intra-actions in ways I didn't . . . I thought I was gonna start doing that once they [her committee] told me, "look at this theory. Read this book." But I'm actually using it to write my mini proposal, which is really interesting. It's helping me in ways I hadn't expected.*

AMY: I'm curious. How is it informing you? What surprises are you encountering as you go through this process?

GINA: *Well, one of the major questions in my proposal is; "What is my relationship with this puppet? What does the Monseñor Romero puppet and performance mean to me?" Even though I'm interviewing folks who were there, and it's a historical question, I had completely overlooked the fact that it's also about me. Intra-actioning is about finding meaning. It starts to feel like, "Yeah, this is the methodology." And thinking of it as a methodology. Not just something that I'm gonna do on the side, but also, something I'm gonna be doing moving forward even after this class[our independent study].*

AMY: I'm wondering, at any point, have you gone back and looked at your own onto-epistemological dispositions? Because that impacts aspects of your thinking; "this is about me", and this relationship you have with the puppet. How you orient yourself as a being in this world is going to impact how you ontologically encounter the puppet, so to speak. So I don't know if you've done that, or linked to that at all.

GINA: *Yes. I'm doing that now. I didn't even think about them till I started interacting with the picture, and pulling things from it that were all relevant to who I am, and my identity as a woman of color, as a theater artist, as an activist. Part of my dissertation is gonna be some type of personal narrative of my journey. I feel like I'm just starting to tap into those things and realize how significant they are, and how it's part of telling that story. That was my first intra-action with the image. It just started to tap into my memory, my sense memory, my emotion, and my consciousness. It just started to open up windows and doors that I think I had disregarded. I was so invested in these other folks in the narrative, and not thinking about how had I not had that experience, I wouldn't be here getting my PhD.*

AMY: Did you map any of that out visually?

GINA: *I have on paper where things are starting to get connected. So, how do I tell the story? What is that story? What was that? Because these events happened almost over a decade ago.*

AMY: You are embedded in between two cultures. Then you also have the in-between of time and place. There's a lot of interesting in-between to contemplate here. One of the things that popped out for me really strongly when I was reading this most recent intra-action, is the idea of thing power. From Jackson and Mazzei, *Thinking with an Agentic Assemblage in Posthuman Inquiry*. Thing power really embodies that idea of object agency and its capacity to exist. It has its own mattering and meaning prior to our encounter, but in our encounter with it, something happens in that in-between. It becomes sort of a relational exchange that makes it, and it makes you. The exchange is the inbetween, it is what actually does the mattering or the making of the experience as opposed to either you or the object.

GINA: *I know there's definitely some literature that I can look into that talks about thing-power when it comes to puppets. And we're looking at a large-scale protest puppet.*

AMY: You're thinking theoretically through these lenses. I think you can work with aspects of different theory and kinda create your own assemblage of theory and practice. We talked about visual thinking strategies, or VTS, last time and the three questions are, "What's going on in this picture?" The second one, the most critical one, asks, "What do you see that makes you say that?" that forces evidentiary reasoning on your part. You're doing that a bit already. But if you sit with that, if you sit in it a little bit and think, "Well, what did I see that made me say that?" Is it the setting? Is it the colors? Is it the facial expression? Is it an experience that you've had? So, is it something I thought rather than something I see? You're already intuitively doing that. Now do it with an attentive awareness, I think this will help connect the little tendrils of how you might tie this together, entangle them in a way that it makes sense. I noticed you said in a couple of places, "*these are silly little things*," and I was curious about this. I thought they were really significant things. What about the scale of the puppet and the ways his hands are outstretched? I invite you to interrogate that. What did you mean by that? Was there a slight discomfort? I wasn't sure if there was something operating under the radar there that, maybe, you still feel that tug towards a different way of knowing or thinking. You're noticing all these other things and you're wanting to give voice. I found all those things very significant and powerful. Even noticing how you think about your thinking, and how you're valuing certain observations more than others, and if so, why? I felt really honored as someone on the outside, to get a little peek into, not only the puppet and what it is, but your own experience with it. What if the puppet responded to its experience of you in first person? If that puppet was going to talk about you 15 years ago (if you can remember back then) what would the puppet notice, see, say? What would the puppet say today? You could perform it!

GINA: *Oh, yeah. Romero and Me. [laughter] And talking about creating this kind of narrative/or play of my encounter with . . . And giving voice to Romero. So in a way, it becomes this kind of magical realism story, fiction non-fiction, that he's kind of communicating with me. So using some of what I experienced,*

AMY: You could just have a conversation and pretend he's your research partner.

GINA: *I was already thinking like, "Okay, this would be a really interesting creative approach to tell this story through a play."*

AMY: You can call him "My Romero."

GINA: *My Romero, yeah. [chuckle]*

AMY: [laughter] Me and my Romero. The 3 Ms, M and M.

GINA: *So he's being canonized this Sunday. He'll be an official saint.*

AMY: Will that impact his voice in your collaboration?

GINA: *I don't know. We'll have to see.*

AMY: Because we are talking about so many theories and methodologies, another interesting exercise/task/architecture you might start is to ask yourself; ontologically what are those things that you truly believe about what 'being' is. I'm imagining satellites of theory, methodology, thing power, and ethnography orbiting around your beliefs, so you have these satellites around your ontology. Do those work together or not? Sometimes they do, sometimes they don't. It's important to know the ontological disposition of a theory because you may be translating it one way, but it may mean something totally different.

GINA: *Yeah.*

AMY: You have to understand how it already lives in the world a little bit and ask; is it the right think-tool for you? Every time you encounter something, where does it fit in this map? And the map itself becomes somewhat rhizomatic. I imagine it as creating a seed of possibility instead of a methodology where you do step one, step two, step three. It's plugging into those realizations from your own ontological disposition, who you are, how you are, and then what you do with that and it becomes its own thing. . . . Your ability to express yourself in only writing is not a reflection of your capacity to think and know. It's not.

GINA: *Yeah. But it's something that's still kind of in you and it's hard not to think about that.*

. . . That's why I wanted to do this with you because I feel like for me, it was an additional tool. It helped . . . It changed my perspective and my confidence as an academic. . . . I talk about the Artistic Research class in my experience and I know other people who are visual thinkers like me, we get excited about that because that's maybe something they struggle with too. My identity as a creative artist, I'm very confident and humble about that work. But when it comes to my identity as an academic or scholar, that's where I'm like, "Well . . . "Trying . . .

Blog Contemplation #3

This is my first intra-action with one of the several photographs I took while at the Pine Forest Theatre in the summer of 2016. I randomly selected a picture and allowed faith or destiny to take its course which was somewhat exciting, especially since I was no longer in charge of what image I would intra-act with next. This picture in particular revealed a number of key themes, concepts, and ideas that I had stored away somewhere in my memory bank . . . I realized how other than a researcher and apprentice

Figure 7.2 Intra-action with PHOTOGRAPH #1: picture of guest circus act from Africa wearing frog masks

I would often perform as well. So many roles to take on! One of the best revelations from this intra-action was the quality of work from our guest performers in comparison to our work as apprentices . . . it was part of the B&P aesthetic and core values that we did not over rehearse and that being messy, disorderly, and chaotic was part of the charm of the theater. This is an important point to consider when talking about B&P and the work of Teatro Miaz (pseudonym), especially when you take into account that poor people's theater or social justice theater tends to prioritize the collaborative spirit of making theater over the final performance. I wonder if this is why B&P speaks to so many other groups who have fewer resources but can still pour their hearts out in a performance?

It is interesting to note here that the intra-action produces more questions rather than answers. It offers multiple entry points into considering her experience. She is also sinking into a highly immersive process much the way we explored ideas and data in the Artistic Research class, slowing down and engaging through multiple senses.

For this intra-action I examined a picture that revealed what it meant to socialize on the farm. An observation from my intra-action is the

Figure 7.3 Intra-action with PHOTOGRAPH #2: picture of apprentice and staff chilling in the front porch

Figure 7.4 Intra-action with PHOTOGRAPH #3: conversation between three images, collage of Esteli (stilt walking), canta-estoria, and making paper maché.

generational gap that is evident from the image. This intra-action made me think of the lack of diversity on the farm and the events that occurred during the apprenticeship (primarily police shootings of unarmed Black and brown people) and how much folks struggled with creating a circus piece that spoke to those issues . . . I am finding intra-actions to be quite fruitful as I am remembering all sorts of key concepts, themes, ideas, and events that took place. Using the images is helping refresh my memory.

I decided to pull three images and see what came of that during my intra-action . . . During this intra-action I realized that if I eliminated my central focus I would be able to uncover everything else happening in the background . . . Coincidentally, I was able to create an entirely new image using the cut outs and the original pictures that represented these themes of labor, hierarchy, and giving and taking orders.

AMY'S RESPONSE ON THE BLOG:

What you are doing here is really quite awesome! One of the things I find interesting about these intra-actions is they seem to unearth marginalized memories that were a part of the tapestry but perhaps a quieter part of the narrative? A part that holds the focus of the B&P. Does this make it less

Figure 7.5 Intra-action with photograph #4: picture of Mongolian horses we work for 4th of July parade.

significant? No. All these aspects of the narrative entangle meaning. How do they shape experiences and perceptions? Were they there quietly informing all along? Was this an example in a subtle way of an invisible intra-action with the B&P and the surrounding community?

Discussion #2

AMY: I see what you're up to [with her proposal]. You talk about using auto-ethnography, phenomenology, oral history, and historiographic methodologies, in addition to artography.

GINA: *Yeah.*

AMY: So what I'm wondering is how do you conceptualize, even though you're kind of separating them somewhat in the chapters, how do you envision them working together or plugging into each other to flesh out what you're up to, with your questions?

GINA: *Plugging into each other. [chuckle]*

AMY: Well, I'm wondering if you're going to find that all of them co-exist in every chapter, but it's like a/r/tography and the Artist-Researcher-Teacher identities. Maybe one or two are emphasized or have more voice while the others are still there in collaboration offering their input . . ..artography by its very nature can be phenomenological.

GINA: *Yeah. What is more present, what's the stronger voice, the artist, the researcher, the educator? I hadn't really thought about that. I do believe that that will happen organically, but I think it's helpful to think about.*

AMY: Keep your awareness on it, noticing. It's that attention thing. Where's your attention going? Is that the lens that really is going to give you the most powerful voice? Or do you need to bring a couple in there and create a choral ensemble of what's being shared?

GINA: *I think that that's one of the discoveries in doing, in intra-acting with the pictures, where things just started to come out organically, and concepts and themes and questions and memory. The storytelling of what that was. But also deeper, not just who I was looking for but what was my experience as a performer, which I had completely forgotten about. I was just thinking about being there as a researcher. I feel like the visual images are really pulling gems. I'm digging for gold and I'm finding out things and connecting them. I think when I write my blog, I'll be able to think a little deeper and I'm going to listen to whatever commentary I made, not just look at the image itself.*

AMY: Did you notice . . . 'Cause now you're juxtaposing different images together. They're still from the same place and even though they're different spaces it's still from a larger common space.

GINA: *Yeah. Yeah.*

AMY: So it's multiples. Like a multiplicity of images. Did you notice your noticing shifted in any way when you were combining them or did you read them differently? Did you read them as a whole or did you read them as three different identities but contextually similar? I'm wondering how your mind made sense of what you were seeing. Did you draw parallels between them or did you make assumptions that this was happening as a result of that or in conjunction with it? I'm just curious. What newness arose from going through this exercise? What was revealed or emerged?

GINA: *That's a good question. I think one of the things is just realizing how many layers there are when I think about space. . . . How each space kind of shifted my identity and the roles that we played in each of the spaces. In some way they're all kind of interconnected. I also think about the power of images and being able to . . . It's hard for me to explain what Pine Forest was to anyone unless you were there. With the photographs, having that kind of visual data to backup the ideas was important. I was focusing on the interviews and prioritizing the text, forgetting that this [images]is how you communicate too. That looking at this image, and just continuing to engage with it, until I felt like I had nothing more to say, but then still looking at it, turning it over, cutting something out, doodling around it, things just started to come up.*

AMY: Staying with it [time and being with]. There's so much conditioning that I think this process has to permeate through those little spaces where there are opportunities to get beyond that; I must find an answer. I must sound smart, or I must use this language or qualifier, validator, correlator, or any of these things that as academics we are habituated to. Do you notice feeling tension as you're in that waiting space, or has that evolved some?

GINA: *I think that's evolved. I feel more confident intra-acting with the work than if I was just doing traditional research. I think it's really changed my perspective of who I am and that I will have something to offer and bring to the table. That has shifted. I still am critical of myself, obviously, but not in the same sense that I had been before I took your class . . . I feel like I'm getting hooked on this visual mapping thing. Again, it goes back to when I was taking your class and thinking, "Yeah, I am the visual thinker. Why am I not thinking this way?" In my day-to-day work I'm doing what I'm now conditioned to do. So this, I feel like, brings me back to my roots . . .*

Blog Contemplation #4

My final assignment for this [independent study] class is to revisit some of the articles on Artography from Artistic Research last semester and to

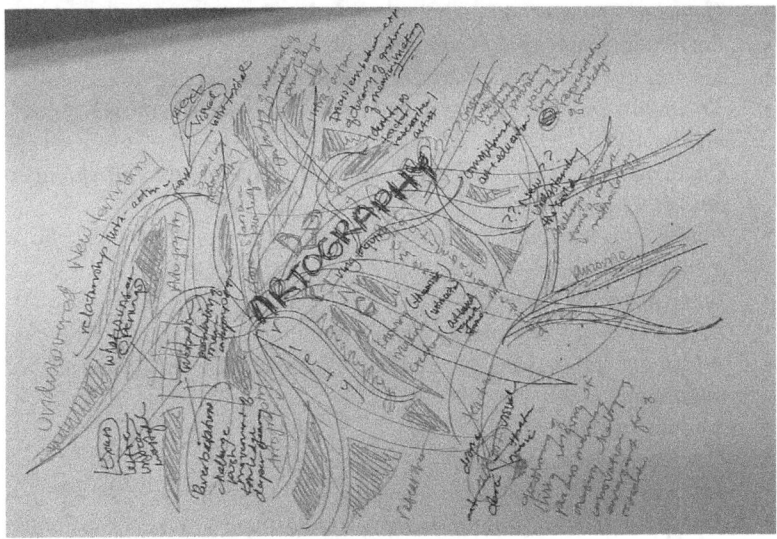

Figure 7.6 Intra-action with "A/R/Tography"

use that information to create a definition of the methodology that I can use throughout my dissertation research. At the very least, begin a draft of my understanding of artography based on these readings and the work throughout the semester. I decided to read the article from last semester that I felt most useful in understanding Artography and intra-act with that article's definition of Artography through a video just like I had done with all the other topics, themes, and materials used in this class. I felt that the best way to understand Artography was to practice Artography. In this case I used Artography to break down any insecurities I might have had about not knowing the methodology and also as a means to process the information I did know and had just reviewed. It was a little bit of both.

My goal is to review the video and jot down some of the main ideas that developed from the intra-action with the reading and use this information to write a clear definition for my final. Below are a few of the findings and information I investigated through my intra-action:
[This was edited down for brevity]
Artography is . . .

- *A sense of doing*
- *When you are doing this doing (1.knowing 2.making 3.creating)*
- *Knowing = theorizing*

- *A form of living inquiry*
- *The why? The constant questioning/challenging of information*
- *Provoking/poking of information*
- *Not just representation of knowledge but actually opposite*
- *Understanding questioning is important as a new way of understanding the world*
- *Challenging traditional forms of research methodologies (providing an answer to a question)*
- *Not just one way but a variety of different ways of doing Artography as a practice*
- *Engaging/embodying*
- *Of material, of ideas, and of knowledge*
- *All about processing the embodied experience of discovery of questioning*
- *Meaning making*
- *Our identity (Artographer)*
- *The Teacher*
- *The Artist (sense of questioning of living inquiry, process, meaning making, innovative, emergent form of research, visual art, theatre, dance, music. Etc. this is the way that we do work already)*
- *The Researcher*
- *Grasping that knowledge is rhizomatic, not linear, but stems from one thing to the next to the next like a root*
- *Active stance on knowledge*
- *Going beyond seeing to doing*
- *About touch (having a relationship intra-action with the work)*
- *Text, visual, intra-textual*
- *Folds*
- *Discovering what is unseen*
- *Reverberation (challenge pushing of knowledge)*
- *Excess (The unseen, the wasteful)*
- *Metaphor (opening possibility of meaning interpretation)*
- *Encounters of material*
- *Embodying it*
- *Questioning it*
- *Experiencing it*
- *All of the in-between that is in between*
- *In between the information, not just looking at knowledge, this is what we know*
- *What the in-between is trying to tell us*
- *Questioning how is it connected?*
- *The investigation to fill in the gaps spaces that are missing*

- *New territory*
- *Undiscovered*

Below is the first draft of my understanding of artography. My goal was to create an introduction to the methodology that those who were unfamiliar with method could easily understand. This is not the end but the beginning of more conversations about Artography.

 Artography is a practice-based research method. Unlike a traditional research approach, which results in a representation of knowledge or in the answering of a specific question, Artography seeks to continuously question and challenge information through direct engagement with the material. This material can consist of texts, images, concepts, and/or ideas. In Artography there is an immediate focus on the engagement, or intra-actions, with the material that allows the researcher not only to encounter the work and its meanings but to embody the work and its significance as well. By embodying the material, the researcher can uncover different ways of learning about the world. The term A/R/Tography incorporates the multi-layered identities and praxis of the Artist, Researcher and Teacher, and the ways these identities coincide or speak with one another. Artography, like the creative process of the "artist," stems from the sense of continuous discovery, towards meaning making through living inquiry and play. According to Rita L. Irwin and Stephanie Springgay in "Artography as Practice-Based Research", living inquiry is "a lifetime commitment to the arts and education through acts of inquiry" (xxix).

 The artographer, like the artist and teacher, is not confined to one way of engaging with material but instead participates in an emergent form of research through non-linear, complex, and rhizomatic ways of meaning-making. Artography, therefore, encourages the artist/researcher/teacher to engage with the material by paying attention to the in-between, or openings of the material. These openings are undiscovered areas we would likely ignore through more traditional research methodologies. Also, the in-between spaces can become folds, after folds, after folds of undiscovered territory. The experience of meaning-making through Artography consists of what Irwing and Springgay call "knowing (theoria), doing (praxis) and making (poesis)" (xxiii).

Final Discussion

GINA: *It was just natural . . . To use the method, to talk about the method.*
AMY: . . . *reading through your list, this was generated after watching the video?*
GINA: *Yes. I feel like an artographer now, versus being introduced to the method. I've been doing it . . . finding ways to incorporate it into my*

own research and my kind of day-to-day learning. I'm much closer to it . . .

AMY: You're putting it in a language that's accessible to those who don't think that way, or maybe use terminology in that way. I would encourage you to maybe develop the idea of the rhizome a little bit more.

GINA: *Yeah, I feel like that's definitely something that's not fulfilled yet . . . even living inquiry, I think I could dig deeper into that as well.*

AMY: Remember when we did our artist statements in class? You could almost take some of these blog entries and cut 'em up and start rearranging them and see what kind of actual statement it makes, and then filling in the blanks as you go . . . I was doing that in my head as I was reading through this, and I'm like, "God, this is such a good list, she's so amazing." This could also describe the rhizome because you have a lot of statements in here that speak to what the rhizome is and what it does. That might give you some of the language that's accessible to explain it. Thinking about your committee specifically, the investigation to fill the gaps and spaces that are missing are like the spaces between the web of the rhizome. It gives them something to visualize. So you create a story, a visual story of what that structure is, what it does, and how it contains emergent information or knowledge. I am totally thrilled and pleased with how articulate you've become about this process. I hope that you are!

GINA: *I did feel and obviously it's not a polished draft but I feel . . . Trusting myself . . . "Okay, you know this. You just read a really good article. You're doing good work." I always wondered what that tension was. I feel like it is part of the constant insecurity, or seeking or finding my academic voice, and I feel like this type of work . . . I feel like I belong in this type because it's familiar to me. I just didn't realize that it was and that it's academic work. And I think as an artist, even as a doctorate student in theater, people constantly just assume that I'm doing something fun. When I tell people I'm getting a PhD in theater, they always ask me, "What are you doing instead of dissertation?" And I say, "A dissertation". [chuckle] We don't just get a free pass. If people knew how hard people in the arts and education work. . . . I'm really glad that we are culminating this with that assignment through the very method itself . . . [chuckle].*

AMY: It's a unique process that's pulling in all of these different tools and assembling them in such a way that you can interrogate what you're up to in a completely different way right from the start. You started in January of last year, so now it really has been a year. You did a semester as a student, now you're an emerging scholar!

References

Holbrook, T., & Pourchier, N. M. (2014). Collage as analysis: Remixing in the crisis of doubt. *Qualitative Inquiry*, *20*, 754–763.

Irwin, R. L., & Springgay, S. (2008). A/r/tography as practice-based research. In M. Cahnmann-Taylor & R. Siegesmund (Eds.), *Arts-based research in education: Foundations for practice* (pp. 103–124). Routledge.

∞ AND ANd And and

Because there is no beginning and no end, I have no conclusion to offer. You should come to your own conclusions or pursue inquiry which might propel your thinking into the new. I offer a simple invitation to pose questions with disruption in mind. Ask messy questions! Probe the edges of your wonderings, and fearlessly engage your curiosities. All things that are, were made up by someone or something.

What are you doing with your attention?
What if you blended aspects of different theories to explain differently?
How can contemplating this ignite exploration in you?
How might you dialog with and (in)between literacies?
What lurks in the margins of your perception?
What verbs propel you to the margins and back again?
What if you could only think in images?
What if there were no words?
What if your attention lingered (in)between the known and unknown?
What if there are no findings, only more questions?
What if . . .?
Why not . . .?
Go ahead . . .. Give yourself permission . . .

DOI: 10.4324/9780367823818-8

Index